Fluids and Electrolytes

A thorough guide covering fluids, electrolytes and acid-base balance of the human body

David Andersson

© Copyright David Andersson 2018-All rights reserved.

The contents of this book may not be reproduced, duplicated, or transmitted without direct written permission from the author.

Under no circumstances will any legal responsibility or blame be held against the publisher for any reparation, damages, or monetary loss due to the information herein, either directly or indirectly.

Legal Notice:

This book is copyright protected. This is only for personal use. You cannot amend, distribute, sell, use, quote, or paraphrase any part or the content within this book without the consent of the author.

Disclaimer Notice:

Please note the information contained within this document is for educational purposes only. Every attempt has been made to provide accurate, up-to-date, and reliable complete information. No warranties of any kind are expressed or implied. Readers acknowledge that the author is not engaging in the rendering of legal, financial, medical, or professional advice. The content of this book has been derived from various sources. Please consult a licensed professional before attempting any techniques outlined in this book.

By reading this document, the reader agrees that under no circumstances is the author responsible for any losses, direct or indirect, which incurred as a result of the use of information contained within this document, including, but not limited to errors, omissions, or inaccuracies.

Contents

Introduction 5

Chapter 1: Distribution of Body Fluids and Homeostasis 7

Chapter 2: Fluid Volume Excess 19

Chapter 3: Fluid Volume Deficit 31

Chapter 4: Acid-Base Homeostasis 43

Chapter 5: Disorders of Acid-Base Balance:
Acidosis and Alkalosis 55

Chapter 6: Sodium: Homeostasis and Imbalances 69

Chapter 7: Chloride: Homeostasis and Imbalances 81

Chapter 8: Potassium: Homeostasis and Imbalances 89

Chapter 9: Calcium: Homeostasis and Imbalances 99

Chapter 10: Magnesium: Homeostasis and Imbalances 109

Chapter 11: Phosphorous: Homeostatis and Imbalances 119

Chapter 12: Imbalances in Common Medical Conditions 129

Conclusion 141

Answers to Exercises 143

Join Our Community 149

Kindle MatchBook 151

Check Out Our Other Books 152

Introduction

Fluid maintenance remains one of the main foundations of medicine. Fluids are deemed as being the most essential substance of life. Around 60% of our body is made up of water, and this serves as a medium for transport of both nutrients and waste products. The electrolytes contained within the body water are responsible for the body's most basic functions, such as nerve function, and functioning of voluntary and involuntary muscles, activating enzymes, and release of hormones. Maintenance of these components in a constant balance is essential, because the entire metabolic process of the body depends on these components. Fluids and electrolytes are not static, but interact continuously with each other, and with other components of the body. It is essential to understand these interactions, and the role they play in maintenance of health.

Imbalances in these substances within the body are responsible for some of the most significant medical conditions and disorders. Any patient, who has presented with a fairly serious medical illness, is bound to have some kind of imbalance in the fluid-electrolyte levels. All patients in intensive care units are routinely screened for such imbalances. Thus, it is apparent that a medical professional needs to have a firm grasped of the fundamentals of electrolytes, fluids and acid-base balance in the body, so that effective treatment can be carried out.

Fluids and Electrolytes

This subject is often confusing for both the beginner medical student and the day to day medical practitioner. However, once the underlying basic concepts of fluids and electrolytes are grasped, it becomes easy to recognize imbalances in these systems, and it follows that treatment becomes easier and more planned.

In this book, we will discuss the regulation of fluids, electrolytes, and the acid-base system, and how these factors are interconnected. When there is an imbalance in one of these electrolytes, the others are usually affected. Similarly, electrolyte imbalances can arise from, or be a result of, acid-base disorders. We will also discuss the signs and symptoms of various imbalances, and touch upon the management modalities for these imbalances. Common medical conditions where there are massive imbalances are described in the last chapter. There are summaries at the end of each chapter, that help you quickly glance through essential information. Lastly, this book tests what you have learned with post chapter tests.

Chapter 1:

Distribution of Body Fluids and Homeostasis

The concept of homeostasis

The term homeostasis is defined as the maintenance of nearly constant conditions in the internal environment. This concept was originally described by the French physiologist, Claud Bernard, who stated that a free and independent existence was possible only because of the stability of the internal milieu. It was Walter Canon in 1960 who coined the term 'homeostasis'.

Most homeostatic mechanisms work by a negative feedback mechanism. This mechanism consists of a sensor, which detects the normal range of values, and an error detector, which detects if the values become abnormal. This then sends a signal to the controller, which interprets the abnormality and stimulates an effector, which works to bring values back into the normal range.

This book discusses homeostasis of fluids, acid-base balance, and electrolytes in the subsequent sections.

Body fluids in homeostasis

The human body is largely made up of water, which exists in the form of several fluids. These fluids play an important role in day to day functioning of the body. Fluids help in temperature regulation, provide protection to vital organs, and lubrication to joint spaces. They also act as a medium to dissolve solutes, and transport essential compounds throughout the body. While water is the solvent, body fluids contain several electrolytes, proteins, and cells, which have to be maintained in homeostasis.

Distribution of body fluids

A large proportion of the human body is made up of fluids. Fluids account for around 50% of total body weight in women and for around 60% of total body weight in men. The proportion of fluids decreases with age, and infants and children generally have a higher percentage of fluid. Fluid proportion also decreases with increase in body fat.

The fluids in the body are distributed between two compartments – intracellular and extracellular. Intracellular fluids (ICF) account for two-thirds of the total content, while extracellular fluids (ECF) account for the remaining one third. The extracellular fluid compartment is further divided into the intravascular and interstitial compartments. The intravascular compartment is the fluid contained in the blood vessels (plasma) and has one-fourth of the total ECF. Interstitial fluid is the fluid that is present between the cells and forms the remaining three-fourths of ECF.

The distribution of fluid between various body compartments is outlined in Figure 1.

Chapter 1: Distribution of Body Fluids and Homeostasis

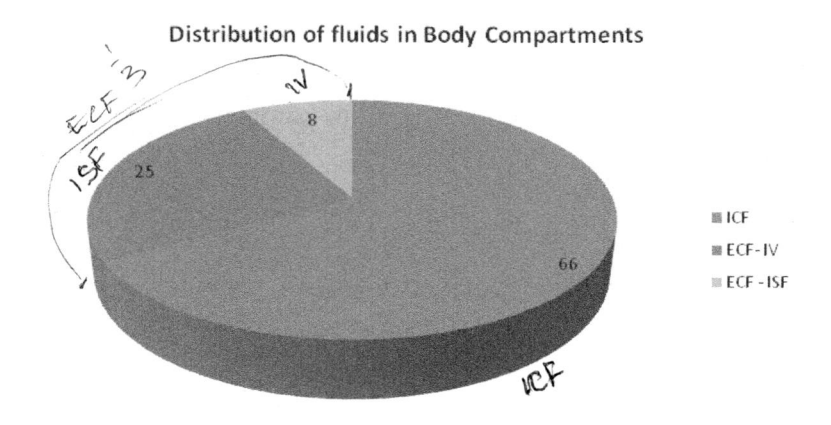

Distribution of fluids in Body Compartments

[Pie chart showing: ICF 66, ISF 25, IV 8, with handwritten annotations: ECF, ISF, ICF]

- ICF
- ECF-IV
- ECF-ISF

Intracellular fluid:

The fluid inside the cell is referred to as cytosol. It contains various enzymes, proteins, and ions. The intracellular fluid contains large quantities of potassium and phosphate ions, and moderate quantities of magnesium and sulfate ions. Sodium, chloride, and calcium ions are present in low quantities. This is in contrast to the ions of the extracellular fluid, which is discussed below.

Extracellular fluid:

The extracellular fluid consists of the interstitial fluid and plasma. The interstitial fluid bathes and surrounds the cells of the body. They aid in transport of nutrients to the cell and waste removal from the cell. The plasma acts as a carrier for the various blood cells and proteins. The ionic composition of the ISF and plasma is similar, however, they differ in the protein composition. The ECF has large amounts of sodium, chloride, and bicarbonate ions. The other ions, which are high in the cytosol, are present only in small quantities here. The plasma contains various dissolved proteins including albumins, globulins, and fibrinogens. It also contains clotting factors and glucose. The interstitial fluid is poorly permeable to proteins and large molecules.

There is another fluid compartment, called the transcellular fluid compartment, which forms 2.5% of the total body fluid content. This does not form a separate fraction of the extracellular fluid, but is a unique part of it. This fluid has protective and lubricative functions. This is located in specific anatomical areas as follows:

- Synovial fluid – Joints
- Cerebrospinal fluid – Brain
- Peritoneal fluid and pleural fluid – Chest cavity, around the layers lining the heart and lungs respectively
- Vitreous humor – Eyeballs

The values of important components of the body fluids are summarized in Table 1.

TABLE 1. CONCENTRATIONS OF ELECTROLYTES IN BODY FLUIDS			
COMPONENT	INTERSTITIAL FLUID (mOsm/L)	PLASMA (mOsm/L)	INTRACELLULAR FLUID (mOsm/L)
Sodium	139	142	14
Chloride	108	108	4
Potassium	4	4.2	140
Calcium	1.2	1.3	0
Bicarbonate	28.3	24	10
Phosphorous	2	2	11
Protein	0.2	1.2	4

Fluid movement between body compartments:

Both the intracellular fluid and extracellular fluid are bound by membranes. These membranes act as a barrier to diffusion of fluids from one compartment to another. While these membranes do not easily allow the passage of solutes from one side of the membrane to another, they allow the passage of water, they are hence referred to as selectively permeable membranes. The process of diffusion of water from a region of high water concentration to an area of low water concentration is referred to as osmosis. The concentration of water in any solution would depend on the amount of solutes that are dissolved in that solution. The total number of solutes (electrolytes, proteins, and other molecules) in a solution are measured in units called 'osmoles', and their proportion in solution is expressed in terms of 'osmolality'. A fluid with greater osmolality is referred to as hypertonic, and a fluid with lesser osmolality is referred to as hypotonic. As plasma has higher quantity of proteins, it tends to have higher osmolality as compared to interstitial fluid.

The interstitial fluid and plasma are separated by the walls of the blood vessels. While the walls of the arteries and veins are thicker, capillaries tend to have thin walls that allow easier movement of water across them. The movement of water across capillaries is governed by two forces – Hydrostatic pressure and Osmotic pressure. Hydrostatic pressure is basically blood pressure, that is, the pressure generated by contractions of the heart during systole. It tends to drive fluid out of blood vessels. Osmotic pressure is the pressure due to the hypertonicity of blood, which tends to retain fluid within vessels.

At the arterial end of the capillary, hydrostatic pressure is always greater. This causes fluid to move out of the blood vessels into the interstitial compartment. At the venous end of the same capillary, hydrostatic pressure decreases and the osmotic pressure is greater. This causes most of the fluid to get reabsorbed into the capillary. About 10% of the fluid fails to get reabsorbed, and this is directed into the lymphatic system, which eventually returns it into circulation.

Factors that regulate fluid balance in the body:

The daily turnover of fluids in the body is around 2300 ml. A human loses this amount of fluid every day. Fluid losses occur largely through urine, and to a lesser extent through feces. Fluids are also lost through evaporation from the skin surface, and through air expelled from the lungs. This is an ongoing process, and is referred to as insensible loss, as the patient is not aware of the loss. Insensible skin loss is diffusion of water from the skin surface and occurs independent of sweating. These fluids must be therefore replenished daily. While most of the fluid input is from dietary sources (food and water), a small amount is also produced from metabolic pathways. This has been summarized in Table 2.

TABLE 2. WATER TURNOVER IN THE HUMAN BODY				
WATER INPUT		WATER OUTPUT	
Method	Quantity (ml)	Method	Quantity (ml)
Ingested food and Water (variable)	2000	Urine (variable)	1100
Metabolic activity	300	Feces	100
		Sweat	100
		Insensible loss (lungs and skin)	700
Total	2300 ml	Total	2300 ml

To maintain the fluid balance, a person needs to take in as much water that leaves the body. If there is an imbalance between input and output, various mechanisms in the body can correct it. The proportion of fluid present in the body is reflected by the plasma osmolality, which is basically the ratio of solutes to solvent. If the fluid (solvent) proportion decreases, there is an increase in plasma osmolality, and vice-versa. This change in osmolality is picked up by osmoreceptors, which are located in the hypothalamus. The hypothalamus then follows two pathways to correct this change:

- Thirst mechanism: If the plasma osmolality increases, the osmoreceptors stimulate the thirst center in the hypothalamus. This triggers a subject to drink more water, thereby decreasing the osmolality of plasma.

Chapter 1: Distribution of Body Fluids and Homeostasis

Antidiuretic hormone pathway: The hypothalamus controls the release of a hormone called the anti-diuretic hormone (ADH) from the posterior pituitary gland. The release of this hormone follows a negative feedback loop. If the plasma osmolality is high, ADH release is stimulated. ADH then acts on the kidneys, causing them to reabsorb water from the distal convoluted tubules and collecting duct. ADH acts on these systems by means of water channel proteins, called aquaporins, which are moved from within the cell to the cell surface. The cell thus becomes more permeable to water, and increased quantities are absorbed. Since less water is excreted, plasma osmolality decreases. On the other hand, if the plasma osmolality decreases, release of ADH is suppressed, and more water gets excreted via the kidneys. This causes the plasma osmolality to rise again.

During increased plasma osmolality, ADH release also causes peripheral arteriole constriction. This ensures that the blood flow is directed towards the core of the body.

Factors that cause fluid imbalance

Certain factors can cause fluid imbalance in the body. Fluid imbalance may be classified into three types as follows:

- Hypovolemia: Decrease in the total body fluid is referred to as hypovolemia. It may occur in the following conditions:
    - Trauma: Due to blood loss
    - Dehydration: Excessive fluid loss, which can occur in vomiting and diarrhea
    - Polyuria – excessive urination, which may be due to drugs like diuretics
    - High fever – increases insensible losses
- Hypervolemia: This is associated with fluid overload. It may be iatrogenic, while treating hypovolemia, or may be secondary to impaired excretion, as in kidney failure.

- Normovolemia with fluid imbalance: This is seen secondary to fluid resuscitation after major loss. In this situation, fluid is retained within the intravascular compartment. There is decreased tissue perfusion, which may lead to multiorgan failure.

SUMMARY:

- Fluids make up 50-60% of the body content, and are essential for normal functioning of the body.
- Fluids in the body are divided between two main compartments – the intracellular and extracellular compartments. The extracellular compartment consists of interstitial fluid and plasma.
- Fluids have a regular turnover in the body and move freely between compartments.
- Fluid balance is regulated and maintained by the hypothalamus, and the antidiuretic hormone.
- Fluid imbalance may occur in certain diseases and conditions.

Chapter 1: Distribution of Body Fluids and Homeostasis

REFERENCES:

1. Guyton, A. C., & Hall, J. E. 1. (2006). Textbook of medical physiology (11th ed.). Philadelphia : New Delhi: Elsevier Saunders.
2. Cooper SJ. From Claude Bernard to Walter Cannon. Emergence of the concept of homeostasis. Appetite 2008; 51: 419-427.
3. Modell H, Cliff W, Michael J, McFarland J, Wenderoth MP, Wright A. A physiologist's view of homeostasis. Advances in Physiology Education. 2015;39(4):259-266.
4. Guyton, A. C., & Hall, J. E. (2000). The body fluid compartments: extracellular and intracellular fluids; interstitial fluid and edema. Textbook of medical physiology, 11, 293.
5. Kreimeier U. Pathophysiology of fluid imbalance. Critical Care. 2000;4(Suppl 2):S3-S7.

Fluids and Electrolytes

EXERCISES:

1. Which population subgroup has a higher percentage of body fluid?
   a. Adult female
   b. Adult male
   c. Infants
   d. Obese people

2. Which of the following is not part of the extracellular compartment?
   a. Interstitial fluid
   b. Synovial fluid
   c. Cytosol
   d. Plasma

3. Which of the following contributes to maximum excretion of fluid from the body?
   a. Sweat
   b. Urine
   c. Feces
   d. Expiration

4. At the arterial end of a capillary, which of the following statements is true?
   a. Hydrostatic pressure > Osmotic pressure
   b. Hydrostatic pressure = Osmotic pressure
   c. Hydrostatic pressure < Osmotic pressure
   d. None of the above

Chapter 1: Distribution of Body Fluids and Homeostasis

5. Which part of the body does insensible loss of fluid occur from?
    a. Kidney
    b. Lungs
    c. Skin
    d. Both b and c

6. On which area of the kidney does the antidiuretic hormone effect its action?
    a. Glomerulus
    b. Proximal collecting tubule
    c. Distal collecting tubule
    d. Excretory duct

7. Which of the following is not an example of transcellular fluid?
    a. Cerebrospinal fluid
    b. Pleural fluid
    c. Renal tubular fluid
    d. Synovial fluid

8. What is the amount of fluid produced by metabolic processes in the body?
    a. 300 ml
    b. 500 ml
    c. 700 ml
    d. 900 ml

Fluids and Electrolytes

9. Which of the following ions is not found predominantly in the intracellular region?

    a. Potassium

    b. Magnesium

    c. Sodium

    d. Phosphorous

10. Which of the following factors tends to excrete fluid from the body?

    a. Thirst mechanism

    b. Increased ADH secretion

    c. Decreased ADH secretion

    d. None of the above

Chapter 2:

Fluid Volume Excess

Fluid volume excess is also known as hypervolemia. This can occur when the volume of fluid intake is far in excess of the fluid that can be excreted from the body. The body has several mechanisms for coping with increased fluid intake. It is only when these mechanisms become saturated, which happens in advanced disease conditions that the clinical manifestation of hypervolemia sets in.

Causes of hypervolemia:

There are two main reasons for which fluid content in the body would increase:

Increased intake: Excessive intake rarely leads to hypervolemia as the body automatically tends to excrete excess oral intake. However, this can occur iatrogenically if excess fluid is given intravenously during fluid resuscitation therapy.

Decreased output:
If the body fails to clear fluids at the same rate as intake, water retention can occur, leading to hypervolemia. This can occur in two ways:

- Failure of excretion of total fluid:

Fluids and Electrolytes

- *[margin note: — Transudative edema →]* This can occur if ① the kidney fails to function, and urine does not form. The kidney retains water, which will later on move to the interstitial space. This is also seen in ② mineralocorticoid excess, which causes sodium and water retention by the kidneys.
- Failure of fluid movement between compartments:
- In the previous chapter, movement of fluids between the capillaries and interstitial spaces was discussed. If any disease process disturbs this movement, fluids may accumulate in the interstitial spaces. This can happen in the following situations:
    - Decreased arteriolar resistance: This may be seen if a patient is taking vasodilator drugs, or there is insufficiency of the sympathetic nervous system. This increases the hydrostatic pressure to drive blood into the interstitial spaces.
    - Increased venous pressure: This may be seen in cardiac failure, in venous obstruction, or in cases where the valvular mechanism of the veins fails, thereby causing backing up of blood from the veins to the capillaries. The backed up blood volume increases the hydrostatic pressure of the capillaries and causes fluid shift into the interstitial space.
    - Decreased plasma osmotic pressure: This can occur if the solutes of the plasma decrease. Plasma protein loss can occur in nephritic syndrome and in burn injuries. The production of plasma proteins may be halted in conditions like liver cirrhosis and severe malnutrition. When the plasma osmotic pressure is low, the capillaries fail to reabsorb the interstitial fluid. This can cause fluid accumulation in the interstitial space. ✓
    - Increased capillary permeability: Inflammatory reactions, immune reactions, ischemia, and infections

[margin note: exudative edema]

can cause capillaries to become 'leaky' which causes fluid escape from them to the interstitial space.

- Lymphatic blockage: A small percentage of interstitial fluid is drained by lymphatics, and, if these channels are blocked, fluid can accumulate over time. This can occur in infectious conditions such as filariasis, which cause lymph node swelling. It can also occur when lymph nodes are surgically removed after treatment for malignant disease.

The interstitial space has various coping mechanisms to deal with the excess fluid overload. One mechanism is by increasing the hydrostatic pressure of the interstitium, so that it resists the capillary pressure and does not receive fluid. The second mechanism is by increasing the lymphatic outflow. Lastly, in later stages, proteins get washed out from interstitial spaces, reducing their osmotic pressure. Fluid accumulation occurs only after these coping mechanisms reach their limits. Therefore clinical manifestations of hypervolemia generally is a sign of advanced disease.

Classification and clinical manifestations of hypervolemia:

When there is fluid volume excess, the water accumulates in tissues just beneath the skin. This type of condition is referred to as edema. Edema usually occurs in dependent areas of the body. The edema may be of two types – pitting or non-pitting. In pitting edema, if pressure is applied to the edematous area, the indentation persists even after pressure is removed. This usually occurs when the interstitial space had fluid but low concentration of protein. In non-pitting edema, the indentation does not persist.

The location of edema may give a clue as to the underlying cause of fluid retention. Based on the location, the various types of edema are as follows:

Fluids and Electrolytes

Generalized edema:

The edema occurs in multiple organs as well as the periphery. This is often seen in cardiac failure, nephrotic syndrome, and severe nutritional deficit. Ascites is a special type of edema that occurs in liver cirrhosis, and involves accumulation of fluid in the peritoneal cavity. Severe generalized edema is referred to as anasarca.

Organ specific edema:

In pathology or inflammation of specific organs, water retention will occur in that organ alone. A few examples are detailed in the table below:

TABLE 1. EXAMPLES OF ORGAN SPECIFIC EDEMA	
Organs affected	Conditions in which this is likely to occur
Cerebral edema – brain	Lack of oxygenation to brain.
Pulmonary edema – lungs	Left ventricular failure, inhalation of toxic agents
Cutaneous edema – skin	Inflammatory reactions e.g. insect bites, myxedema
Periorbital edema	Trauma to the orbit
Lymphedema	Areas where lymph drainage is blocked e.g. elephantiasis of the foot in filariasis.

The clinical features of water retention will also depend on the area where edema occurs. Most cases of edema present with swelling of the involved region. Since the skin over the area is stretched, there may also be pain in that region. Limitation of motion of the affected limb may also occur. Organ specific edema will have symptoms related to that organ.

- Peripheral edema presents with swelling of the ankles and wrists.

Chapter 2: Fluid Volume Excess

- Pulmonary edema presents with shortness of breath.
- Cerebral edema presents with drowsiness, or loss of consciousness.
- Effusions: In organ specific edema, fluid can go into potential spaces such as pleura or pericardium. Pleural effusion presents with shortness of breath, and pericardial effusion has breathlessness in addition to muffled heart sounds.

Diagnostic tests:

If a patient presents with unexplained edema, a series of tests may be performed to diagnose or rule out systemic diseases. These include the following:

- Brain natriuretic peptide measurement: for congestive cardiac failure
- Blood urea nitrogen, creatinine, and urine analysis : for renal disease
- Liver enzymes and albumin measurement: for liver disease
- Ultrasonography, D-dimer tests: when deep vein thrombosis and venous insufficiency of lower limbs is suspected.

Management of hypervolemia:

The underlying cause of hypervolemia must be identified and managed appropriately. Once the disease process comes down, edema tends to resolve. Certain measures, however, can be taken to help reduce the amount of water retention. These are as follows:

- Proper positioning: Elevating swollen legs causes fluid to drain from them as it works against gravity.
- Intermittent pneumatic compression devices: These devices provide external pressure that forces blood and lymph out of that area.
- Limiting sodium intake in diet may help reduce water retention.

- Diuretics: these are drugs that promote excretion of water from the kidneys. They help reduce the overall water content in the body. They are useful in cardiac and liver disease, but must not be used in conditions of increased capillary permeability.

SUMMARY:

- Fluid retention or hypervolemia occurs if there is increased intake or decreased output.
- Decreased output is more common and occurs when the kidneys fail, or interstitial fluid fails to move into capillaries.
- The main clinical manifestation of hypervolemia is edema. It may be generalized or localized, pitting or non-pitting.
- Edema presents with swelling, pain, limitation of motion, and organ specific features.
- Resolving the underlying cause can remove edema. Other measures include positioning, pressure devices, and diuretic drugs.

REFERENCES:

1. Braunwald E, Loscalzo J. Edema. In: Longo DL, Fauci AS, Kasper DL, Hauser SL, Jameson JL, Loscalzo J, eds. Harrison's Principles of Internal Medicine. 2011 18th ed. New York, NY: McGraw-Hill
2. Trayes KP, Studdiford JS, Pickle S, Tully AS. Edema: diagnosis and management. Am Fam Physician. 2013 Jul 15;88(2).
3. O'Brien JG, Chennubhotla SA, Chennubhotla RV. Treatment of edema. Am Fam Physician. 2005;71(11):2111-2117
4. Rockson SG. Lymphedema. Am J Med. 2001;110(4):288-295

Fluids and Electrolytes

EXERCISES:

1. Which one of these options causes water retention?

    a. Heart attack

    b. High blood pressure

    c. Kidney Failure

    d. All the above

2. Which of these mechanisms does not cause water retention?

    a. Increased plasma osmotic pressure

    b. Reduced plasma osmotic pressure

    c. Increased capillary permeability

    d. Increased tissue osmotic pressure

3. What electrolyte contributes to increased water retention?

    a. Potassium

    b. Magnesium

    c. Sodium

    d. Calcium

Chapter 2: Fluid Volume Excess

4. Which of the following is not a symptom of edema?

    a. Swelling

    b. Pain

    c. Dizziness

    d. Limitation of motion

5. Ascites is a type of edema that occurs in which one of the following conditions?

    a. Renal failure

    b. Liver failure

    c. Cardiac failure

    d. Pulmonary failure

6. Which of the following is not a coping mechanism that the interstitial fluid follows in fluid overload?

    a. Increasing the hydrostatic pressure of the interstitium

    b. Increasing the lymphatic outflow

    c. Decreasing the osmotic pressure of the interstitium

    d. Increasing the hydrostatic pressure of the plasma

Fluids and Electrolytes

7. Elephantiasis is a type of edema that is seen in which type of infection?

    a. Malaria

    b. Dengue

    c. Filariasis

    d. Amoebiasis

8. What is the term used to refer to generalized edema?

    a. Amyloidosis

    b. Anasarca

    c. Lymphedema

    d. None of the above

9. Pericardial effusion is characterized by which of the following clinical features?

    a. Increase heart sounds

    b. Muffled heart sounds

    c. Irregular heart sounds

    d. All of the above

Chapter 2: Fluid Volume Excess

10. In which of the following causes of fluid overload are diuretics contraindicated?

    a. Renal failure

    b. Increased capillary permeability

    c. Heart failure

    d. None of the above

Chapter 3:

Fluid Volume Deficit

A fluid volume deficit can happen if the person loses more fluids from his body than he takes in. Fluid loss can occur from any of the compartments. Fluid loss that occurs exclusively from the intravascular space is referred to as hypovolemia. On the other hand, dehydration is a form of fluid deficit that is spread across all the fluid compartments.

DEHYDRATION:

This is a deficit of total body water. There is loss of fluid from all the body compartments.

Causes:
Dehydration can occur in two scenarios – if the total water intake has decreased, or if there is excessive total water loss from the body.

- Decreased intake:
    - Blunted thirst mechanism – seen in elderly patients
    - Lack of available water – This can occur in extreme situations, such as when a person is lost in the desert, or at sea.

Fluids and Electrolytes

- When a patient is asked to be 'nil per mouth' before surgery, and adequate fluids are not given intravenously.
- Excessive water loss:
  - Vomiting and diarrhea can quickly cause loss of large volumes of water.
  - Excessive sweating: can occur with sports, overexertion, and in high fevers.
  - Excessive urination: In uncontrolled diabetics, and with drugs such as diuretics.

Classification of dehydration:

Generally, when body fluid is lost, there is variable loss of electrolytes as well, particularly sodium. Depending on the amount of electrolyte that is lost, dehydration can be classified as follows:

TABLE 1. TYPES OF DEHYDRATION		
Type of dehydration	Plasma osmolarity	Condition seen
Isonatremic	130 - 150 mEq/L	Equal loss of sodium and water
Hyponatremic	< 130 mEq/L	Lost fluid has excessive sodium
Hypernatremic	> 150 mEq/L	Lost fluid has minimal sodium

Signs and symptoms:

The signs and symptoms depend on the stage of dehydration in which the patient presents.

Chapter 3: Fluid Volume Deficit

Mild dehydration (fluid loss of 30-50 ml/kg)

- The earliest symptom is a feeling of dryness of mouth and tongue.
- Headache, tiredness, and loss of appetite
- Reduced urination (unless the cause is polyuria). This is a sign that can be checked in infants, who will have fewer wet diapers.
- Infants and children will also present with irritability
- Capillary refill time is 2 seconds
- Heart rate is normal or increased. Blood pressure and respiratory rate are normal.

Moderate dehydration: (fluid loss of 60-100 ml/kg)

- Dryness of mouth and mucous membranes increases
- Sunken face and eyes
- Skin is dry and inelastic, turgor is slow
- There will be reduced lacrimation, along with decreased urine output.
- Dizziness and lethargy may be present.
- Capillary refill time: 2-4 seconds
- Heart rate increases, blood pressure drops and respiratory rate increases.

Severe dehydration: (fluid loss of 90-150 ml/kg)

- The features resemble that of moderate dehydration, but are increased in severity.
- Patient will be disoriented and confused.
- Skin and mucous membranes become parched and cracked. Skin turgor is lost completely and tenting takes place.
- Capillary refill time is more than 4 seconds.
- Urinary output is very low; anuria may be present

Fluids and Electrolytes

Consequences of severe dehydration:

Seizures: This can occur if the plasma sodium falls.

Pontine necrosis and cerebral edema: This is a consequence of rapid rehydration, and not the dehydration process itself. Rapid rehydration causes rapid rush of water into cells, which may rupture, leading to cerebral edema.

Management of dehydration:

Mild dehydration: This can be corrected by rehydrating the patient with water alone. The patient can take sips of water, or ice chips to correct this state.

Moderate dehydration: Usually this involves loss of water as well as electrolytes. Oral rehydration solution (ORS) is indicated for replacement of water as well as lost electrolytes such as sodium and potassium. It is particularly useful in malnourished children from developing countries, who are affected with diarrhea. If oral intake is ineffective because of repeated vomiting, replacement can be done via nasogastric tube.

Severe dehydration: This requires prompt replacement of fluids, but at a slow rate to avoid the complications mentioned above. Intravenous or intraosseous lines must be established. Fluid replacement is preferably done with isotonic solutions such as Ringer's lactate or 0.9% sodium chloride. If there is symptomatic hyponatremia, replacement is done with slow infusion of 3% saline solution. Repeated monitoring is essential to adjust the amount of fluid given.

Supportive therapy and treating the cause of dehydration are important steps in management. Anti-emetics such as Ondansetron have been found effective in severe vomiting. However, routine use of antibiotic therapy and anti-diarrheal agents is not recommended.

Chapter 3: Fluid Volume Deficit

HYPOVOLEMIA:

In hypovolemia, fluid is lost from the intravascular compartment alone. It is also called volume depletion. This can occur when whole blood is lost (e.g. during trauma or surgery), or when plasma alone is lost (e.g. in burn injuries). It can also occur as a consequence of dehydration, when excessive sodium is lost from the body.

The most important consequence of hypovolemia is that the body goes into a shock response in order to sustain life and function. This is called hypovolemic shock. Since loss of whole blood is the commonest cause, it is also referred to as hemorrhagic shock.

Causes:
- External blood/plasma loss: Traumatic injuries, burns
- Internal blood loss: internal organ injury, rupture of aneurysms, bleeding gastrointestinal ulcers.
- Pregnancy related disorders : ectopic pregnancy, placenta previa, placenta abruption
- Extreme fluid loss in diarrhea, vomiting etc.

Signs and symptoms:
The clinical presentation of shock depends on the severity, which in turn depends on the volume of blood that is lost.

Class I:
- Less than 15% of blood volume is lost
- No symptoms
- Blood pressure, pulse rate, and respiratory rate are normal. Mild tachycardia may be seen.
- Capillary refill time is about 3 seconds.

Class II:
- Blood volume loss is 15 – 30%

Fluids and Electrolytes

- Clinically manifested by pallor and anxiety
- Tachycardia and tachypnoea is seen. Blood pressure may be increased slightly.

Class III:

- Blood volume loss is 30 - 40%
- Mental status is altered; confusion and agitation may be seen.
- Oliguria is seen
- Marked tachycardia and tachypnoea is present. Systolic blood pressure begins to fall.

Class IV:

- Blood volume loss is more than 40%.
- Depression of mental status and loss of consciousness.
- Skin is pale and cold.
- Minimal or no urination.
- Extreme fall in systolic blood pressure; tachycardia is seen.

Management of hypovolemic shock:

Hypovolemic shock is a life threatening condition and therefore must be managed promptly both in the field and in the emergency department. There are three basic goals of management:

- Maximizing oxygen delivery to the tissues:
    - Airway must be cleared of obstruction
    - Oxygen must be delivered at high flow to improve the concentration of oxygen being delivered.
- Fluid resuscitation:
    - Two wide bore intravenous lines must be started for fluid resuscitation. Intraosseous lines can be used in

Chapter 3: Fluid Volume Deficit

children and intra-arterial lines may be used in severe cases.

- o Initially, fluid bolus is given with crystalloids. A bolus of 1 – 2 liters is given in adults and 20 ml/kg is given for children. Colloid solutions are not preferred for initial resuscitation.
- o Further fluids are given based on response assessment; in the meantime, patient's blood is sent for typing and cross-matching in case further blood transfusions are required.
- o If there is no response to the fluid bolus, or if the patient presents with Class IV shock, type O-negative blood must be given.

- Controlling ongoing blood losses:
  - o Ongoing blood losses must be controlled after obtaining intravenous access. Direct pressure must be applied on areas of active bleeding in the field; bleeding vessels can be identified and ligated in the emergency room.
  - o Fractures of long bones must be reduced and splinted.
  - o Bleeding from nose and other cavities must be controlled by packing.
  - o Surgical intervention may be needed for bleeding control; e.g. removal of spleen.

SUMMARY:

- Fluid deficit may include either fluid loss from all the body compartments (dehydration), or loss from only the intravascular compartment.
- Dehydration occurs either due to decreased intake of water or excessive loss of fluids from the body.
- It presents with dryness of skin and mucous membranes, headache, and lethargy. Later stages may show mental confusion and seizures.
- Dehydration must be treated with oral rehydration solutions or intravenous fluid therapy.
- Hypovolemia is life threatening and can cause shock.
- It is manifested as tachycardia, reduced blood pressure, and mental depression or loss of consciousness.
- The goals in management of hypovolemia include maximizing oxygen delivery to tissues, prompt fluid resuscitation and replacement, and preventing further loss.

Chapter 3: Fluid Volume Deficit

REFERENCES:

1. Dill DB, Costill DL. Calculation of percentage changes in volumes of blood, plasma, and red cells in dehydration. Journal of applied physiology. 1974 Aug;37(2):247-8.

2. Dalal KS, Rajwade D, Suchak R. "Nil per oral after midnight": Is it necessary for clear fluids? Indian Journal of Anaesthesia. 2010;54(5):445-447. doi:10.4103/0019-5049.71044.

3. Colletti JE, Brown KM, Sharieff GQ, Barata IA, Ishimine P, ACEP Pediatric Emergency Medicine Committee. The management of children with gastroenteritis and dehydration in the emergency department. J Emerg Med. 2010 Jun. 38(5):686-98

4. Goldman RD, Friedman JN, Parkin PC. Validation of the clinical dehydration scale for children with acute gastroenteritis. Pediatrics. 2008 Sep. 122(3):545-9

5. Gulati A. Vascular endothelium and hypovolemic shock. Curr Vasc Pharmacol. 2016. 14(2):187-95

6. Reinhart K, Perner A, Sprung CL, et al. Consensus statement of the ESICM task force on colloid volume therapy in critically ill patients. Intensive Care Med. 2012 Mar. 38(3):368-83

Fluids and Electrolytes

EXERCISES:

1. Which is the earliest manifestation of dehydration?
    a. Lethargy
    b. Confusion
    c. Dry mouth
    d. Seizures

2. What is the commonest cause of dehydration in children in developing countries?
    a. Insufficient intake
    b. Fluid shift
    c. Diarrhea
    d. Polyuria

3. What is the ideal treatment for moderate dehydration due to diarrhea in children?
    a. Plain Water
    b. Oral rehydration solution
    c. Intravenous fluids
    d. Glucose drinks

4. Hypovolemia is a deficit of fluid in which compartment?
    a. Extra cellular Fluid
    b. Intra cellular fluid
    c. Interstitial Fluid compartment
    d. Intravascular compartment

Chapter 3: Fluid Volume Deficit

5. Which of the following is not an immediate goal for managing hypovolemia?
    a. Maximizing oxygenation
    b. Reducing neurological deficit
    c. Controlling ongoing blood loss
    d. Fluid replacement

6. Mild hypovolemia involves loss of how much quantity of fluid?
    a. less than 5%
    b. Less than 10%
    c. Less than 15%
    d. Less than 20%

7. What is the possible consequence of rapid intravenous rehydration?
    a. Cerebral edema
    b. Pulmonary edema
    c. Cardiac ededma
    d. Generalized edema

8. What is the clinical presentation of vitals seen in hypovolemic shock?
    a. Decreased pulse and blood pressure
    b. Increased pulse and blood pressure
    c. Decreased pulse, increased blood pressure
    d. Increased pulse, decreased blood pressure.

Fluids and Electrolytes

9. In which stage of shock does the mental status alter?

    a. Stage I

    b. Stage II

    c. Stage III

    d. Stage IV

10. What is the initial fluid preferred for resuscitation in hypovolemic shock?

    a. Crystalloids

    b. Colloids

    c. Whole blood

    d. Packed cells

Chapter 4:

Acid–Base Homeostasis

While the volume of fluids in the body must be maintained in perfect balance, its composition and properties, such as pH, must also be maintained in harmony. The homeostasis of individual electrolytes will be discussed in future chapters. In this chapter, the regulation of pH, or maintenance of the body's acid-base balance is discussed in detail.

The body maintains the pH if the intracellular fluids and extracellular fluids remain at a constant level. ECF is slightly basic, and ranges from 7.35 to 7.45, usually it is maintained optimally at 7.40. This optimum pH provides an ideal environment for components of the fluid to function. The immune system can function only at this optimal pH, and above or below this pH, there is a tendency for proteins to get denatured, leading to loss of function.

There are three main methods by which the body regulates acid-base homeostasis:

- Chemical buffering mechanisms
- Renal regulation of pH
- Respiratory regulation of pH

Fluids and Electrolytes

Chemical buffering systems:

Chemical buffers are compounds that bring back the pH of body fluids to their normal value, thereby helping to maintain homeostasis. They are the first line of defense against sudden changes in pH. These buffers are basically molecules or compounds that are capable of combining with hydrogen ions, or releasing them as per the need. There are three important chemical buffers present in the human body:

Bicarbonate buffer:

- The bicarbonate buffer is the most important chemical buffer system and is largely responsible for maintaining the pH of the extracellular fluid.
- Bicarbonate ion (HCO_3^-) can combine with hydrogen ion (H^+) to form carbonic acid (H_2CO_3). This can be represented by the equation below:
- $HCO_3^- + H^+ \longleftrightarrow H_2CO_3$
- When the pH of the ECF falls, it contains excess H^+ ions, which combine with HCO_3. Therefore the above equation shifts to the right. The carbonic acid further dissociates into carbon dioxide and water (CO_2 and H_2o), and excess carbon dioxide is removed by the lungs.
- When the pH of the ECF rises, there is a deficit of H^+ ions. So carbonic acid dissociates, and the above equation shifts to the left.
- At normal pH, the usual ratio of bicarbonate to carbonic acid is 20:1.
- The bicarbonate buffer system depends on two factors: Firstly, normal levels of bicarbonate must be present in the body. Secondly, the respiratory system must function properly. If it does not, carbon dioxide levels may be abnormal which would affect the efficacy of this system. The bicarbonate system and respiratory regulation are interlinked, as the

Chapter 4: Acid-Base Homeostasis

carbon dioxide produced affects respiratory regulation. This can be seen in the next section.

Protein buffer:

- This buffer mechanism acts both intracellularly and extracellularly.
- Proteins are made up of amino acids. Each amino acid contains a carboxyl group (COO-) and an amino group (NH2+).
- Both the carboxyl and amino groups are capable of adding on another hydrogen ion when needed, and can transform into COOH and NH3+ under acidic conditions.
- These hydrogen ions can be released from the amino acids when the pH increases.

Phosphate buffer:

- Phosphate plays an important role in buffering of intracellular fluid.
- Phosphate ions (HPO4-) can take up excess hydrogen ions, under conditions of acidity, to become dihydrogen phosphate (H2PO4) or trihydrogen phosphate (H3PO4).
- They release these ions when the pH increases.

Respiratory regulation of pH:

The rate of respiration is one of the mechanisms of maintaining acid-base balance in the body.

In general, the pH of the blood, and content of carbon dioxide (pCO2) in the blood and extracellular fluid, can be sensed by central chemoreceptors. These chemoreceptors are located in the medulla oblongata of the brain.

Fluids and Electrolytes

If there is any change in the pH or pCO2, this is picked up by the chemoreceptors, which then send signals to the respiratory center, located in the pons and medulla oblongata.

The respiratory center sends efferent signals via motor neurons to the respiratory muscle, namely the diaphragm and intercostal muscles. This will alter the rate of contraction of these muscles, which in turn will alter the rate of ventilation.

Normally, the pCO2 (partial pressure of carbon dioxide) is 40 mmHg, and alveolar ventilation allows about 15 mol of CO2 to be excreted each day.

If the acidity of blood rises (low pH), the pCO2 increases. This causes increase in alveolar ventilation, which will blow off more carbon dioxide, thereby decreasing the pCO2 level.

On the other hand, if the blood becomes more alkaline (high pH), the pCO2 falls. This causes the alveolar ventilation to decrease, which retains CO2 in the blood, and allowing the pCO2 to rise.

It must be remembered, from the previous section, that CO2 in the blood is a product of carbonic acid breakdown, and is formed in response to low pH. Therefore high pCO2 reflects low pH and vice versa.

Renal regulation of pH:

The kidneys play an extremely important role in acid-base balance, because, they are capable of taking up the slack if the other systems stop functioning for any reason. There are two ways in which the kidneys regulate acid-base balance:

- Reabsorption of bicarbonate into the body
- Excretion of acids into the urine.

Chapter 4: Acid-Base Homeostasis

Reabsorption of bicarbonate:

- Almost all the bicarbonate present in the plasma is filtered by the glomerulus, which is roughly 4000 to 5000 mmol/day.
- However, this is later reabsorbed. 85 - 90% of the bicarbonate is reabsorbed in the proximal tubule, while the rest gets reabsorbed in the intercalated cells of the distal convoluted tubule and collecting duct.
- Within the tubular cell, water and carbon dioxide, in the presence of the enzyme carbonic anhydrase, combine to form bicarbonate and hydrogen ions:
- $H_2O + CO_2 \leftrightarrow H^+ + HCO_3^-$
- The hydrogen ion leaves the tubular cell and enters the lumen, using Na+H+ antiporter, which exchanges H+ for Na+.
- Within the lumen, the hydrogen ion combines with the filtered bicarbonate ion, once again forming carbon dioxide and water.
- Carbon dioxide, which is lipid soluble, enters the cell again, and combines with water to produce bicarbonate and hydrogen ion. Thus, while bicarbonate is reabsorbed, hydrogen ion is not lost; instead, there is a net gain of one sodium ion.

Excretion of acids:

There are two types of acids which are excreted

Ammonium (NH4+): Ammonia (NH3) is produced as a breakdown product of glutamine, which is present in the tubular filtrate, as well as the peritubular capillaries. This tends to combine with hydrogen ions and is excreted in the collecting duct as ammonium. This serves to reduce the pH of the final urine as well.

Titrable acids: Titrable acids include phosphates, citrates, and sulfates. These compounds have the capacity to combine with hydrogen ions and get excreted from the kidney, thereby reducing the overall pH.

SUMMARY:

- For optimum functioning of the body, the pH of extracellular fluid has to be maintained at 7.35 to 7.45.
- This is done by three mechanisms – chemical buffering systems, respiratory mechanisms, and renal mechanisms.
- Chemical buffers include bicarbonates, phosphates, and protein buffers. These compounds are capable of binding with or dissociating from hydrogen ions as the need arises.
- Respiratory system regulates pH by controlling the amount of carbon dioxide that is exhaled. In acidic states, excess CO_2 is exhaled, and in basic states, CO_2 is retained to maintain the pH.
- Renal system regulates pH by two methods- firstly, by absorbing bicarbonate back from the renal tubular fluid, and secondly, by excretion of ammonium and other acids, that removes excess hydrogen ions from the body.

REFERENCES:

1. Hamm, L. L., Nakhoul, N., & Hering-Smith, K. S. (2015). Acid-Base Homeostasis. Clinical Journal of the American Society of Nephrology : CJASN, 10(12), 2232-2242.

2. Boyd, D. (1974). Acid base homeostasis and its disorders. Chest, 66(2), p.25.

3. Hood VL, Tannen RL. (1998). Protection of acid-base balance by pH regulation of acid production. N Engl J Med 339: 819-826.

4. Tannen RL. (1983). Ammonia and acid-base homeostasis. Med Clin North Am; 67(4): 781-98.

Fluids and Electrolytes

EXERCISES:

1. What is the optimum pH of human blood?

    a. 7.20

    b. 7.30

    c. 7.40

    d. 7.50

2. Which of the following is not a component of the chemical buffering system?

    a. Ammonium

    b. Bicarbonate

    c. Protein

    d. Phosphate

3. What is the normal ratio of bicarbonate to carbonic acid in the body?

    a. 10:1

    b. 30:1

    c. 20:1

    d. 40:1

Chapter 4: Acid–Base Homeostasis

4. The efficacy of the bicarbonate buffer depends on proper functioning of which of the following systems?

    a. Renal system

    b. Cardiac system

    c. Nervous system

    d. Respiratory system

5. Which component of the amino acid acts as a chemical buffer?

    a. R group

    b. Central carbon atom

    c. Carboxyl group

    d. Hydrogen atom

6. Which of the following buffers has a predominantly intracellular effect?

    a. Ammonium

    b. Phosphate

    c. Bicarbonate

    d. Proteins

Fluids and Electrolytes

7. In which part of the body are the chemoreceptors for PaCO2 located?

    a. Lungs

    b. Carotid body

    c. Pons

    d. Medulla oblongata

8. How much carbon dioxide is excreted by the lungs per day under normal conditions?

    a. 5 mol

    b. 10 mol

    c. 15 mol

    d. 20mol

9. In which area of the kidney does most of the bicarbonate reabsorption occur?

    a. Proximal convoluted tubule

    b. Descending lop of Henle

    c. Thick ascending loop of Henle

    d. Distal convoluted tubule

Chapter 4: Acid-Base Homeostasis

10. Which amino acid predominantly contributes to the production of ammonium ion?

    a. Aspartate

    b. Glutamine

    c. Methionine

    d. Arginine

Chapter 5:

Disorders of Acid-Base Balance: Acidosis and Alkalosis

The normal pH of blood is 7.35 to 7.45. If the pH moves beyond these limits, acid-base disorders can result. Acidosis occurs when arterial blood pH moves below 7.35, whereas alkalosis occurs when the pH rises above 7.45.

Depending on the mechanism of homeostasis that is disturbed or overloaded, acid-base disorders may be categorized as respiratory disorders or metabolic disorders. Metabolic disorders are primarily due to changes in levels of bicarbonate within the body, while respiratory disorders are due to changes in the partial pressure of carbon dioxide.

Having either a metabolic or a respiratory acid-base disorder is referred to as a 'simple' disorder. There is another category of acid-base disorders, namely the mixed disorders, which have both metabolic and respiratory components. This may be because the body is trying to compensate for one type of disorder. However, it usually occurs independently. Here both metabolic and respiratory disorders can co-exist because of independent pathological processes in the disease. Such conditions can lead to extreme changes in pH and are dangerous.

TESTS FOR DIAGNOSIS OF ACID-BASE DISORDERS:

In order to correctly diagnose the type of acid-base disorder, certain tests must be carried out in a step-wise manner. These are described as follows:

Arterial blood gases:
- Blood drawn from the arteries is analyzed for pH and PaCO2. From these values, the bicarbonate is calculated using the following equation:
- $pH = 6.1 + \dfrac{\log HCO_3^-}{PaCO_2 \times 0.0301}$
- An increase in pH indicates alkalosis, while a decrease in pH indicates acidosis.
- Respiratory acidosis and alkalosis would be associated with increase and decrease in PaCO2 respectively. Bicarbonates are normal.
- Metabolic acidosis and alkalosis would be associated with a respective decrease and increase in HCO3.
- However, in compensation, the above values may be altered.

TYPE OF ACID BASE DISORDER	pH	pCO2	HCO3-	Compensatory mechanism
Metabolic acidosis	Decreased	Normal	Decreased	Respiratory- reduced pCO2
Metabolic alkalosis	Increased	Normal	Increased	Respiratory- increased pCO2
Respiratory acidosis	Decreased	Increased	Normal	Renal – increased HCO3-
Respiratory alkalosis	Increased	Decreased	Normal	Renal – decreased HCO3-

Chapter 5: Disorders of Acid-Base Balance: Acidosis and Alkalosis

Anion gap:

- Anion gap is defined as the difference between the cations and anions in the blood plasma. It is therefore a measure of the neutrality of plasma.
- The major cation in the blood is sodium, and the major anions in blood are chloride and bicarbonate.
- The anion gap is measured as: Na - (Cl + HCO3).
- The normal anion gap is 6 - 12 mEq/L.
- This gap may increase in there is a decrease in unmeasured anions, and, rarely, due to increase in unmeasured cations. The anion gap is useful in diagnosing type of metabolic acidosis.

METABOLIC ACIDOSIS

This is a condition where there is a relative decrease in serum bicarbonate concentration.

Causes:

The cause can be of two types depending on the anion gap measurement.

Normal anion gap:

In these cases, the anion gap is normal, usually because there is an increase in chloride, which has compensated for the decrease in bicarbonate. Hence this variety is also referred to as hyperchloremic acidosis. This may be due to:

- Gastric losses - vomiting, diarrhoea
- Acidifying drugs - Calcium chloride
- Renal tubular acidosis
- Hyperkalemia - due to drugs, or renal dysfunction

High anion gap:

The anion gap is increased due to reduction in bicarbonate, or increase in acidifyting cations. These are seen in the following conditions:

- Diabetic ketoacidosis
- Ketoacidosis due to alcoholism and starvation
- Lactic acid acidosis
- Renal failure
- Toxins such as ethylene glycol, propylene glycol, methanol, and salicylates.

Signs and symptoms:

Generally the signs and symptoms are non specific. The respiratory center gets stimulated and tries to compensate by removing excess carbon dioxide from the body. Therefore the following signs may be seen:

- Increased ventilation
- Dyspnoea
- Chest pain

Other signs are as follows:

- Palpitations
- Headache, fatigue, and confusion
- Nausea, vomiting, appetite loss

Management:

The management of metabolic acidosis depends upon two factors – the severity and the cause.

Severe acidosis (pH < 7.15), irrespective of the cause, is generally corrected by alkali therapy. This involves intravenous administration of sodium bicarbonate. 50 to 100 ml ampules of sodium bicarbonate (containing 50-100 mEq NaHCO3), is added to one litre solution

Chapter 5: Disorders of Acid-Base Balance: Acidosis and Alkalosis

of 5% dextrose or 0.25% normal saline, and is administered over 30 to 45 minutes. Serum electrolytes, must be monitored throughout as alkali therapy can result in hypernatremia or hypokalemia.

The goal of alkali therapy is not to achieve normal pH, but to bring the pH to 7.2. Further correction is better achieved by correction of the underlying cause. Continuous alkali therapy can have adverse effects such as volume overload, and increased $PaCO_2$, especially in patients with respiratory dysfunction. In lactic acidosis, it can even exacerbate the acidosis by stimulating phosphofructokinase, and must therefore be used with caution in these patients.

Cause - based treatment:

- Diabetic ketoacidosis : Insulin therapy.
- Starvation and alcohol related acidosis: Intravenous glucose
- Salicylate toxicity: Gastric lavage with saline, followed by activated charcoal
- Ethylene glycol/ methanol toxicity: Saline osmotic diuresis, antidote such as fomepizole.

METABOLIC ALKALOSIS

This condition is associatedwith an increase in serum bicarbonate levels.This is generally associated with hypochloremia and hypokalemia.

Causes:
Based on the amount of chloride that is lost by the body, it can be categorized as chloride responsive alkalosis and chloride resistant alkalosis.

Causes of chloride responsive alkalosis:

- Loss of gastric HCl - vomiting, nasogastric suctioning
- Ingestion of excessive antacids

- Loop diuretics and thiazide diuretics.

Causes of chlride resistant alkalosis:

- Hyperaldosteronism
- Renin increase – tumors
- Autosomal recessive syndromes – Bartter syndrome, Gitelman syndrome
- Hypokalemia, Hypomagnesemia

Signs and symptoms:

These are non-specific and may include:

- Weakness
- Myalgia
- Arrythmias
- Depression of ventilation – as a compensatory mechanism, to retain carbon dioxide.

Management:

- If the alkalosis is chloride responsive, it may be corrected by intravenous infusion of isotonic sodium chloride. Potassium chloride is preferred if hypokalemia is present or edema is present.
- Vomiting/Gastric suction: Use of H2blockers or proton pump inhibitors to minimize hydrogen ion losses
- Diuretics: These can be stopped or replaced with acetazolamide or potassium sparing diuretics.
- Chloride resistant alkalosis must be treated based on the cause. Aldosteronism and rennin increase due to tumors must be corrected by surgery. Dexamethasone therapy may be used to suppress the HPA axis. Syndromes are corrected with potassium supplementation and ACE inhibitors, and potassium sparing diuretics.

Chapter 5: Disorders of Acid-Base Balance: Acidosis and Alkalosis

- Severe alkalosis (pH> 7.55) may require administration of HCl. Patients in advanced renal failure must additionally undergo dialysis.

RESPIRATORY ACIDOSIS

Respiratory acidosis occurs when there is alveolar hypoventilation. The lungs are unable to excrete the excess carbon dioxide, leading to an increase in PaCO2.

Respiratory acidosis may be of two types – acute and chronic. In acute acidosis, the pH is low and PaCO2 is high. In chronic acidosis, the renal system tries to compensate for the acidity by retaining bicarbonate. This will reflect a normal pH, but causes elevated serum bicarbonate.

Causes:

Acute respiratory acidosis:

- Airway obstruction in acute phases of Asthma or COPD
- Respiratory muscle paralysis: Myasthenia gravis, Guillian Barre syndrome

Chronic respiratory acidosis:

- Obstructive sleep apnoea
- Severe asthma or COPD
- Intersitial fibrosis
- Chest wall deformities

Signs and Symptoms:
- Dyspnoea
- Anxiety
- Disturbed sleep, daytime sleepiness

- As CO2 content of the blood increases - confusion and delirium is seen, called carbon dioxide narcosis.
- Arterial blood gases show hypoxia as well as hypercapnoea, and decreased pH.

Management:

As is the case with other acid-base disorders, treatment is aimed at correcting the underlying respiratory disease. Respiratory ventilation may be improved by the use of:

- Bronchodilators –salbutamol, ipratropium bromide, theophylline etc.
- Respiratory stimulants such as medroxyprogesterone.
- Oxygen supplementation –to correct hypoxia.

In life threatening cases, endotracheal intubation and mechanical positive pressure ventilation must be used to excrete carbon dioxide from the system.

RESPIRATORY ALKALOSIS

This is caused by an increase in alveolar ventilation, which causes excess of carbon dioxide to be excreted from the body. As with acidosis, this can also be acute (high pH) or chronic (normal pH) in which the kidneys have compensated by excreting bicarbonate, leading to low serum levels.

Causes:

- Centrally - due to stimulation of respiratory center:
    - Fear and anxiety (commonest cause)
    - Stroke
    - Head injury
    - Drugs and toxins - progesterone, analeptics etc.
- Pulmonary stimulation - asthma, pulmonary embolism, pneumonia

Chapter 5: Disorders of Acid-Base Balance: Acidosis and Alkalosis

Signs and Symptoms:
- Sweating
- Fast, shallow breathing
- Palpitations, tremors

Management:

Since this condition is usually caused by anxiety, calm reassurance is an important aspect of treatment.

Patients are asked to rebreathe into a paper bag. This causes them to inhale more carbon dioxide, thereby increasing the blood levels.

Patients who do not respond may be administered sedatives or narcotic drugs. Beta adrenergic blockers can help reduce the adrenergic effects such as palpitations.

SUMMARY:

- Any change in the pH of the blood can result in acidosis or alkalosis.
- Metabolic acidosis and alkalosis are the result of either decreased or increased levels of bicarbonate in the body.
- Respiratory acidosis and alkalosis are the result of either increase or decrease in the partial pressure of carbon dioxide in the body.
- As the symptoms of these disorders are non-specific, diagnosis is generally made by arterial blood gases, and anion gap measurement.
- The primary treatment should be aimed at correcting the underlying cause of acidosis or alkalosis.
- Severe metabolic acidosis may be corrected by infusion of sodium bicarbonate, while severe metabolic alkalosis may be corrected by administering hydrochloride.

Fluids and Electrolytes

- Respiratory acidosis is corrected by assisting ventilation and providing positive pressure ventilation.
- Respiratory alkalosis is corrected by rebreathing carbon dioxide and administering sedatives if required.

Chapter 5: Disorders of Acid-Base Balance: Acidosis and Alkalosis

REFERENCES:

1. Reddy P, Mooradian AD. Clinical utility of anion gap in deciphering acid-base disorders. Int J Clin Pract. 2009 Oct. 63(10):1516-25

2. Effros RM, Wesson JA. Acid-Base Balance. Mason RJ, Broaddus VC, Murray JF, Nadel JA, eds. Murray and Nadel's Textbook of Respiratory Medicine. 4th ed. Philadelphia, PA: Elsevier Saunders; 2005. Vol 1: 192-93

3. Kraut JA, Madias NE. Lactic Acidosis: Current Treatments and Future Directions. Am J Kidney Dis. 2016 Sep. 68 (3):473-82

4. Gennari FJ. Pathophysiology of metabolic alkalosis: a new classification based on the centrality of stimulated collecting duct ion transport. Am J Kidney Dis. 2011 Oct. 58(4):626-36

5. Cham GW, Tan WP, Earnest A, Soh CH. Clinical predictors of acute respiratory acidosis during exacerbation of asthma and chronic obstructive pulmonary disease. Eur J Emerg Med. 2002 Sep. 9(3):225-32

6. Gardner WN. The pathophysiology of hyperventilation disorders. Chest. 1996 Feb. 109(2):516-34

Fluids and Electrolytes

EXERCISES:

1. A high anion gap is seen in which one of the following conditions?

    a. Metabolic acidosis

    b. Metabolic alkalosis

    c. Respiratory acidosis

    d. Respiratory alkalosis

2. Which value remains normal in uncompensated metabolic acidosis?

    a. pH

    b. Bicarbonate

    c. Partial pressure of carbon dioxide

    d. None of the above

3. Which of the following is a compensatory mechanism for metabolic alkalosis?

    a. Increased bicarbonate

    b. Decreased bicarbonate

    c. Increased pCO2

    d. Decreased pCO2

4. What is the mainstay of treatment in all acid-base disorders?

    a. Infusing sodium bicarbonate

    b. Infusing hydrochloric acid

    c. Wait and observe

    d. Correct the underlying cause

Chapter 5: Disorders of Acid-Base Balance: Acidosis and Alkalosis

5. Metabolic acidosis may be associated with which one of the following electrolyte disturbances?

    a. Hyperchloremia

    b. Hypochloremia

    c. Hypernatremia

    d. Hyponatremia

6. Saline osmotic diuresis is the recommended treatment for which cause of metabolic acidosis?

    a. Diabetic ketoacidosis

    b. Ethylene glycol toxicity

    c. Salicyate toxicity

    d. Lactic acidosis

7. Sodium bicarbonate therapy is given for metabolic acidosis when the pH falls below which level?

    a. 7.25

    b. 7.15

    c. 7.35

    d. 7.45

8. Excessive ingestion of antacids may lead to which one of the following acid base disorders?

    a. Metabolic acidosis

    b. Metabolic alkalosis

    c. Respiratory acidosis

    d. Respiratory alkalosis

Fluids and Electrolytes

9. Daytime sleepiness and confusion is a sign of which of the following abnormalities?

    a. Metabolic acidosis

    b. Metabolic alkalosis

    c. Respiratory acidosis

    d. Respiratory alkalosis

10. What is the recommended treatment for respiratory alkalosis?

    a. Rebreathing oxygen

    b. Rebreathing carbon dioxide

    c. Normal breathing

    d. None of the above

Chapter 6:

Sodium: Homeostasis and Imbalances

As discussed in the earlier chapters, electrolytes form an important component of body fluids. Therefore, fluid regulation does not involve the regulation of water alone, individual electrolytes must also be regulated. The regulation of sodium and chloride, and, to a lesser extent, potassium, is closely interlinked with fluid regulation, in order to maintain osmolarity. Incorrect osmolarity would cause cells to shrink or swell, both of which can damage cellular integrity.

REGULATION OF PLASMA SODIUM:

Sodium has several important roles in the human body. It plays an important role in the regulation of blood pressure. It is important for impulse transmission in proper functioning of nerves and muscles. Since sodium is the main solute in extracellular fluid, its regulation is very important in order to maintain osmolarity. The normal level of sodium in the serum is maintained at 135-145 mEq/L. Regulation of sodium within this range is achieved by several ways, which are outlined below:

The renin-angiotensin system:

- The juxtaglomerular cells of the kidney can detect the sodium concentration of the renal tubular fluid, as it passes through the proximal convoluted tubule, and Henle's loop.
- If there is a lowering of plasma sodium concentration, these cells detect it and release an enzyme called renin into the bloodstream.
- Renin acts on a plasma protein called angiotensinogen, and converts it into angiotensin I.
- The pulmonary capillaries secrete another enzyme, the angiotensin converting enzyme (ACE). When the blood containing angiotensin I passes through the lungs, ACE acts on angiotensin I and converts it into angiotensin II.
- Angiotensin II stimulates the adrenal cortex to release a hormone called aldosterone.
- Aldosterone acts on the distal convoluted tubule and collecting duct of the kidney. It causes exchange of potassium for sodium ions in the renal tubular fluid. Thus sodium ions are reabsorbed and potassium ions are excreted.
- The adrenal cortex can also directly sense plasma osmolarity, and cause direct release of aldosterone when needed.
- Another hormone, called atrial natriuretic peptide (ANP) is released by the heart in high sodium states. It causes increased excretion of sodium by the kidneys.

IMBALANCE OF SODIUM IN THE BODY

HYPONATREMIA:

Hyponatremia occurs when the plasma sodium concentration falls below 135 mEq/L. Depending on the amount of decrease in the sodium concentration, hyponatremia may be mild (130-134 mEq/L), moderate (125-129 mEq/L), or severe (< 125 mEq/L).

Chapter 6: Sodium: Homeostasis and Imbalances

Causes:

- Hyponatremia may or may not be associated with a change in total body water and other solutes. Depending on the osmolarity of the extracellular fluid, hyponatremia may be hypertonic, normotonic, or hypotonic. Depending on the water content, it may be classified as hypovolemic, euvolemic, or hypervolemic hyponatremia.

- Hypertonic hyponatremia: In this, the total body sodium is normal. However, due to the increased presence of other osmotically active solutes, sodium gets forced from the extracellular to intracellular compartment. This may be seen in diabetes (increased glucose), or administration of drugs such as mannitol.

- Normotonic hyponatremia: This is seen in patients who have abnormally high levels of lipids or proteins, which prevents the sodium levels from being measured.

- Hypotonic hyponatremia: This may be of three types:
    - Hypovolemic: Both free water and solute content are reduced. This is seen in:
        - Renal disease: polycystic kidney disease, interstitial nephropathy etc.
        - Cerebral salt wasting disorders: subarachnoid hemorrhage, infectious meningitis, carcinoma etc.
        - Diuretic drugs, particularly thiazide diuretics.
    - Hypervolemic: This is usually associated with clinical edema. Examples: Liver cirrhosis, congestive cardiac failure, nephritic syndrome
    - Normovolemic: This is usually seen in hospitalized patients after major surgery, trauma, or infection. It is associated with syndrome of inappropriate antidiuretic hormone secretion (SIADH).

Fluids and Electrolytes

Signs and symptoms:
- Tiredness, lethargy, and malaise
- Nausea
- Headache
- Depressed consciousness
- Seizures, coma

Management:
Hyponatremia is a serious condition that must be managed promptly to prevent cerebral edema. The treatment must follow guidelines given by the expert committee panel on hyponatremia. These recommendations are outlined below:

- Acute hyponatremia cases (which have a duration of 24-48 hours) must undergo urgent correction.
- For patients who have severe symptoms, who are at higher risk of neurological damage, 100 ml of 3% sodium chloride must be infused intravenously over ten minutes.
- This can be repeated as needed, up to three times.
- In patients with mild to moderate symptoms, the infusion rate may be reduced to 0.5 – 2 ml/kg/hr.
- In cases with a chronic presentation, (duration of more than 48 hours), correction must be done at a slower rate to avoid the osmotic demyelination syndrome (ODS), which can cause necrosis of cells in the pontine region of the brain.
- The infusion rate of such cases must be 4-8 mmol/L in 24 hours.
- In patients with SIADH, treatment includes fluid restriction, increased intake of solutes, and treatment with loop diuretics must be done.
- In hypovolemic cases, fluid requirement must also be balanced.

Chapter 6: Sodium: Homeostasis and Imbalances

HYPERNATREMIA:

Hypernatremia occurs when the plasma sodium concentration rises above 145 mmol/L. Generally this occurs when there is a decrease in the water content of the extracellular fluid relative to the sodium content. Hypernatremia due to true sodium increase is very rare. The hyperosmolarity of the extracellular fluid tends to draw water out of the intracellular compartment, leading to cell shrinkage. This is particularly significant in neuronal cells and can lead to brain damage. The low volume of the ECF (due to water loss) can also have circulatory consequences, such as hypotension and tachycardia.

Causes:
- Factors causing increased loss of water. This is seen in cases of dehydration due to lack of access, increased sweating etc.
- Renal fluid loss in excess of electrolytes: loop diuretic drugs, diabetic osmotic diuresis, acute tubular necrosis.
- Diabetes insipidus: this is a condition of ADH deficiency, in which water cannot be reabsorbed.
- Impaired thirst mechanism

Signs and symptoms:
These are similar to the signs of dehydration discussed previously.
- Cognitive dysfunction – lethargy, confusion, irritability, seizures
- Tachycardia, decreased blood pressure
- Dry skin and mucous membranes

Management of hypernatremia:
- The management of hypernatremia depends on the duration of symptoms.
- Acute hypernatremia (clinical onset of less than 24 hours) must be corrected rapidly.
- Chronic hypernatremia (>24 hours, usual clinical onset greater than 48 hours) requires slow correction. This is be-

cause, in chronic hypernatremia, certain amount of adjustment would have taken place, especially in the brain cells. Water from intracellular compartment would have moved extracellularly to maintain volume, leading to higher intracellular osmotic pressure. In this scenario, if fluids are rapidly infused, the water would rush into the cells, leading to cellular edema and demyelination.

- Therefore, in acute hypernatremia, a rate of 2 - 3 mEq/L/hr is generally used, up to a maximum of 12 mEq/L/day. On the other hand, in chronic hypernatremia, the rate of infusion must be restricted to 0.5 mEq/L/hr, and the total must not exceed 8 - 10 mEq/L/day.

- The amount of fluid to be infused depends on the patient's total body water, and the amount of sodium present in the fluid that is used for infusion.

- Patient's total body water is calculated by multiplying the patient's body weight with the correction factor. Correction factors are listed in Table 1.

TABLE 1. CORRECTION FACTOR VALUES FOR TOTAL BODY WATER		
AGE GROUP	MALE	FEMALE
Children	0.6	0.6
Young adults	0.6	0.5
Elderly	0.5	0.45

- The fluid to be infused is chosen depending on the amount of sodium that it contains. For instance, 5% dextrose contains no sodium, 0.2% sodium chloride contains 34 mmol/L, and 0.45% sodium chloride contains 77 mmol/L.

- The change in serum sodium concentration for one liter of infused fluid is calculated as follows: the difference between infused sodium and serum sodium is calculated. This is divided by total body water+ 1 (1 being the water volume

Chapter 6: Sodium: Homeostasis and Imbalances

added by infusion). This change in sodium concentration can guide the volume of fluid that is to be infused in a single day.

- Although the above method is standard, there is a great deal of interindividual variation. Both the clinical signs and laboratory values must be monitored during treatment. Neurological examination must be done periodically, and if an improvement in clinical signs is seen, the rate of infusion may be reduced. Serum and urine electrolytes must also be assessed every 1-2 hours.
- The cause of hypernatremia must also be determined and corrected simultaneously. Better water access, better control of diabetes mellitus, etc. are some examples. Hypernatremia die to diabetes insipidus must be corrected by synthetic ADH (vasopressin).

SUMMARY:

- Sodium is the most significant electrolyte that contributes to plasma osmolarity. It is also essential for muscle and nerve function.
- Regulation of sodium is largely done through the rennin-angiotensin-aldosterone pathway.
- There are several causes for hyponatremia, and it may be associated with changes in water and other solutes of the extracellular fluid.
- Prompt treatment of hyponatremia must be done to prevent cerebral edema. 3% hypertonic saline is given for treatment.
- Hypernatremia usually occurs with dehydration. It must be corrected by infusion of fluids that are low in sodium.
- For both hyponatremia and hypernatremia, chronic cases must be corrected slowly to prevent irreversible brain damage.

REFERENCES:

1. Fried LF, Palevsky PM. Hyponatremia and hypernatremia. Med Clin North Am. 1997 May. 81(3):585-609

2. Verbalis JG, Goldsmith SR, Greenberg A, Korzelius C, Schrier RW, Sterns RH, et al. Diagnosis, evaluation, and treatment of hyponatremia: expert panel recommendations. Am J Med. 2013 Oct. 126 (10 Suppl 1):S1-42

3. Adrogue HJ, Madias NE. Hyponatremia. N Engl J Med. 2000 May 25. 342(21):1581-9.

4. Hoorn EJ, Zietse R. Diagnosis and Treatment of Hyponatremia: Compilation of the Guidelines. J Am Soc Nephrol. 2017 May. 28 (5):1340-1349

5. Adrogue HJ, Madias NE. Hypernatremia. N Engl J Med. 2000 May 18. 342(20):1493-9.

6. Lindner G, Schwarz C, Kneidinger N, et al. Can we really predict the change in serum sodium levels? An analysis of currently proposed formulae in hypernatraemic patients. Nephrol Dial Transplant. 2008 Nov. 23(11):3501-8

7. Muhsin SA, Mount DB. Diagnosis and treatment of hypernatremia. Best Pract Res Clin Endocrinol Metab. 2016 Mar. 30 (2):189-203

Chapter 6: Sodium: Homeostasis and Imbalances

EXERCISES:

1. What is the most important role of sodium in the body?
    a. It increases bone strength
    b. It is an integral part of the neuromuscular network
    c. It aids the continuous pumping of the heart
    d. It plays a role in acid base balance

2. What hormones affect sodium concentration?
    a. Adrenaline and testosterone
    b. Aldosterone and testosterone
    c. Aldosterone and anti-diuretic hormone
    d. Adrenaline and anti-diuretic hormone

3. What is the earliest sign of sodium deficiency?
    a. Headache
    b. Abdominal pain
    c. Fatigue
    d. Seizures

4. What organ does aldosterone signal to control sodium levels?
    a. Lungs
    b. Skin
    c. Kidneys
    d. Adrenal gland

Fluids and Electrolytes

5. How does our body routinely excrete sodium?
    a. Through the urine
    b. Through bowel movement
    c. Through the saliva
    d. Through the skin

6. Which signs and symptoms are not associated with sodium excess?
    a. Lethargy
    b. Seizures
    c. Confusion
    d. Insomnia

7. What are the consequences of rapid correction of hypernatremia?
    a. Cerebral edema
    b. Pulmonary edema
    c. Generalized edema
    d. Peripheral edema

8. Increased excretion of sodium in the urine is facilitated by which one of the following?
    a. Adrenaline
    b. Anti-diuretic hormone
    c. Atrial natriuretic peptide
    d. Angiotensin

Chapter 6: Sodium: Homeostasis and Imbalances

9. What fluid is used for infusion in patients with hyponatremia?

    a. 0.9% sodium chloride

    b. 0.25% sodium chloride

    c. 2% sodium chloride

    d. 3% sodium chloride

10. Synthetic vasopressin must be used for correction of which of the following causes of hypernatremia?

    a. Renal tubular necrosis

    b. Diabetes mellitus

    c. Diabetes insipidus

    d. Juvenile diabetes

Chapter 7:

Chloride: Homeostasis and Imbalances

Chloride is a negatively charged ion which can combine with cations in the body such as sodium and hydrogen ions. Along with sodium, it plays an important role in maintaining the osmotic pressure of body fluids. It forms a key component of the digestive acid of the stomach (hydrochloric acid). It also interacts with other ions in the body during the regulation of acid base balance. The normal blood level of chloride is 97-108 mmol/l

REGULATION OF CHLORIDE IN THE BODY:

Chloride regulation generally occurs parallel to regulation of sodium. The levels are regulated by balancing dietary intake with the amount that is excreted and reabsorbed by the kidneys. Unlike sodium, reabsorption of chloride is a passive process. In the proximal convoluted tubule, initially there is active reabsorption of sodium and water, followed by various other solutes. This causes an increase of chloride concentration on the tubular fluid as compared to plasma, by the time the fluid reaches the end of the proximal convoluted tubule. Owing to the difference in this concentration, chloride gets reabsorbed into the plasma. However, in the collecting duct, chloride either gets absorbed or secreted in exchange for bicarbonate ions. The amount would be determined by the existing chloride levels in the body.

IMBALANCES:

HYPOCHLOREMIA:

This occurs when the chloride level goes below 97 mmol/L. Hypochloremia can lead to alkalosis, because, chloride ions are generally exchanged for bicarbonates in the renal tubules. In the absence of chloride, bicarbonates are retained, which increase the alkalinity of the blood.

Causes:
- Inadequate dietary intake
- Defective renal tubular absorption
- Excessive loss – vomiting, diarrhea, especially chloride losing diarrhea in children, excessive nasogastric tube suctioning.

Signs and Symptoms:
Hypochloremia generally does not present with specific signs and symptoms. However, in infants who have hypochloremic alkalosis, the following features may be seen:

- Vomiting and watery diarrhea
- Failure to thrive
- Lethargy, confusion, and seizures
- Muscle weakness and cramps
- Abdominal distention

Management:
- Since hypochloremia rarely occurs in isolation, the patient must be assessed for dehydration, hyponatremia, and hypokalemia.

Chapter 7: Chloride: Homeostasis and Imbalances

- In cases which appear to be severe (patient in shock), aggressive fluid resuscitation must be started with isotonic fluids, preferably 0.9% sodium chloride. If hypokalemia is present, potassium chloride may be added.

- Serum electrolytes must be reassessed after 6 hours in order to calculate maintenance.

- If hypochloremia is due to gastrointestinal cause, such as acid reflux or pyloric stenosis, surgical intervention must be done.

- Mild cases of hypochloremia may be corrected by increasing the dietary intake alone.

HYPERCHLOREMIA:

Hyperchloremia occurs when the serum chloride level rises above 107 mmol/L. It is usually associated with acidosis, as, increased levels of chloride are seen only when bicarbonates are deficient.

Causes:
- Dehydration
- Excessive intake of salt/sea water
- Cystic fibrosis
- Congestive cardiac failure
- Metabolic acidosis

Signs and Symptoms:
- Nausea, tiredness
- Mental confusion progressing to stupor
- Tachypnoea may be present if there is an attempt at respiratory compensation.

Management:
The management of hyperchloremia would depend on the cause.

Fluids and Electrolytes

- Excess ingestion must be corrected by withholding further salt intake.
- Metabolic acidosis is corrected by bicarbonate infusion.
- Dehydration is treated by water replenishment.

SUMMARY:

- Chloride is the predominant anion of the extracellular fluid, and along with sodium, maintains the osmolarity of the ECF.
- Chloride homeostasis is passive and parallels sodium homeostasis. The amount of chloride excreted and reabsorbed is passively adjusted based on serum levels.
- Since chloride indirectly affects acid-base balance, hypochloremia can cause alkalosis, and hyperchloremia can cause acidosis.
- Hypochloremia is treated by infusion of isotonic saline.
- Hyperchloremia is treated by water replenishment and withholding further chloride ingestion.

Chapter 7: Chloride: Homeostasis and Imbalances

REFERENCES:

1. Berend, K., van Hulsteijn, L., & Gans, R. (2012). Chloride: The queen of electrolytes?. European Journal Of Internal Medicine, 23(3), 203-211.
2. Galla, J. (1988). Chloride Transport And Disorders Of Acid-Base Balance. Annual Review Of Physiology, 50(1), 141-158.
3. Tani, M., Morimatsu, H., Takatsu, F., & Morita, K. (2012). The Incidence and Prognostic Value of Hypochloremia in Critically Ill Patients. The Scientific World Journal, 2012, 1-7.
4. Al-Abbad A, Nazer H, Sanjad SA, Al-Sabban E. Congenital chloride diarrhea: A single center experience with ten patients. Ann Saudi Med. 1995 Sep. 15(5):466-9
5. Nagami, G. (2016). Hyperchloremia – Why and how. Nefrología, 36(4), 347-353.

Fluids and Electrolytes

EXERCISES:

1. What is the approximate normal value for serum chloride?

    a. 87 -97 mmol/l

    b. 97 – 107 mmol/L

    c. 107 – 117 mmol/ L

    d. 117- 127 mmpl/L

2. Reabsorption of chloride occurs parallel to which other electrolyte?

    a. Potassium

    b. Sodium

    c. Magnesium

    d. Calcium

3. Which of the following is not a treatment for hypochloremia?

    a. Sodium chloride

    b. Potassium chloride

    c. Calcium chloride

    d. Dietary chloride

Chapter 7: Chloride: Homeostasis and Imbalances

4. Which of the following signs is not seen in a hypochloremic patient?

    a. Vomiting

    b. Bone pain

    c. Muscle cramps

    d. Lethargy

5. Which of the following is not a cause for hyperchloremia?

    a. Dehydration

    b. Inadequate dietary intake

    c. Cystic fibrosis

    d. Congestive cardiac failure

Chapter 8:

Potassium: Homeostasis and Imbalances

Potassium is an important electrolyte that plays an important role in multiple functions of the body. It is primarily an intracellular cation. It plays an important role in determining the resting membrane potential of the cell and maintains intracellular osmotic balance. It is crucial for several enzyme activities, and plays an important role in cell growth and division. Besides this, potassium has a key role to play in maintenance of acid-base balance in the body.

REGULATION OF POTASSIUM:

Potassium is acquired by the body primarily though dietary intake. 90% of body potassium is excreted through urine, while the remaining 10% is excreted through feces. Most of the potassium acquired by the body is taken up by the cells, and only 2% remains in the extracellular fluid. However, if there is a decrease in the ECF concentration of potassium for any reason, the cells release potassium in order to maintain this balance.

Renal excretion of potassium occurs at various levels. While potassium is filtered out initially by the glomeruli, about 2/3rd is reabsorbed passively at the level of the proximal convoluted tubule. Some amount of potassium gets secreted in the descending loop of Henle and gets reabsorbed again in the thick ascending loop. However, the final determinant of the amount of potassium to be excreted takes place at the distal convoluted tubule and collecting duct. There is active secretion which is regulated by the hormone aldosterone.

The cells of the zona glomerulosa of the adrenal cortex are sensitive to changes in serum potassium levels. If the serum potassium is high, the adrenal cortex releases aldosterone, which in turn acts on the distal convoluted tubule and collecting duct and causes increased secretion of potassium into the tubular fluid. Similarly, if the serum potassium is low, the secretion of aldosterone and hence excretion of potassium is suppressed.

The normal serum value of potassium is 3.5 - 5.2 mmol/L. The body is extremely sensitive to changes in potassium, and even small changes can have major manifestations. These changes are detailed below.

HYPOKALEMIA

Hypokalemia occurs when the serum potassium level falls below 3.5 mmol/L. Depending on the value of decrease, it may be categorized as mild hypokalemia (3.0 - 3.5 mmol/L), moderate (2.5 - 3.0 mmol/L), or severe hypokalemia (less than 2.5 mmol/L).

Causes:
- Inadequate dietary intake: In elderly patients, patients with eating disorders such as anorexia and bulimia etc.
- Excessive excretion of potassium:
    - Adrenal tumors that produce excess aldosterone
    - Uncontrolled diabetes: Excess glucose in the blood can cause osmotic dieresis

Chapter 8: Potassium: Homeostasis and Imbalances

- o Drugs causing increased excretion: Diuretics, mannitol
- o Gastrointestinal losses: Vomiting, diarrhea.
- o Autosomal recessive disorders associated with increased potassium excretion: e.g. Bartter syndrome, Gitelman syndrome, Gullner syndrome etc.
- Potassium shift: Sometimes potassium can shift from the extracellular to intracellular compartment, causing low serum levels. This may be seen in alkalosis, hypothermia, and refeeding conditions (e.g. starvation), and with some drugs such as insulin and beta-agonists.

Signs and Symptoms:

Patients with mild hypokalemia usually do not show signs and symptoms. There may be some weakness, tiredness, and fatigue, but the patient rarely automatically associates this with low potassium levels.

Most symptoms are related to muscle function and cardiac function. These may be:

- Muscle pain and cramping
- Palpitations
- Severe cases – bradycardia and collapse of the cardiovascular system.
- Neurological signs may be seen in severe cases – hallucinations, delirium, psychosis etc.

Management:

The aim of management of hypokalemia is twofold – identifying the cause for potassium loss, in order to minimize or stop it, and to replenish the body's existing potassium stores.

Stopping the cause:

- Any diuretic, laxative, or other drug that is suspected of causing hypokalemia must be discontinued immediately.

Fluids and Electrolytes

- Potassium sparing diuretics are a feasible alternative.
- For potassium deficit due to gastrointestinal loss, drugs must be administered to stop diarrhea and vomiting.
- Patients who are receiving nasogastric tube suctioning must be placed on H2 blockers.
- In uncontrolled diabetics, glucose levels must be brought under control to prevent potassium loss due to polyuria.
- Surgical intervention may be required to remove the cause in certain cases, such as adrenal adenomas.

Replenishment of potassium:

- In mild and moderate hypokalemia, replacement may be done through oral fluids.
- Severe hypokalemia requires immediate intravenous potassium.
- This must be done at a slow rate of 10 mEq/hour.
- Potassium replacement should be followed closely with ECG monitoring throughout.
- Serum potassium levels should be measured periodically to avoid potential hyperkalemia.

HYPERKALEMIA

Hyperkalemia occurs when the serum potassium rises beyond 5.0 mmol/L. Hyperkalemia is a serious condition that can escalate quickly in severity and have fatal consequences. Hence, it must be identified and managed promptly. When unmanaged, hyperkalemia can lead to cardiac arrest and death.

Causes:

- Increased intake of potassium:

Chapter 8: Potassium: Homeostasis and Imbalances

- - Dietary intake – this is especially common with 'salt substitutes' which contain large quantities of potassium
  - Intravenous infusion: during hypokalemia correction
  - Intravenous infusion of drugs such as penicillin G or packed red cells.
- Decreased excretion of potassium:
  - Renal diseases that affect the distal tubule of kidney. These include diabetic nephropathy, and urinary tract obstruction.
  - Addison's disease and Genetic disorders
- Shift of potassium from the intracellular space into extracellular space: This can occur in cell death such as tumor lysis or muscle lysis, metabolic acidosis, toxicity of chemotherapeutic drugs such as methotrexate and cyclosporine.

Signs and Symptoms:

Just like hypokalemia, there could be little to no symptoms for hyperkalemia. The reported symptoms are usually vague and are related to muscle function or cardiac function. These can include:

- Weakness and fatigue
- Tingling of the skin
- Palpitations and chest pain
- Nausea, vomiting

Any patient who is suspected of having hyperkalemia must have an ECG in addition to serum and urine potassium testing. This is because false positives for hyperkalemia in blood tests are common. Blood cells may rupture, releasing the potassium from the intercellular fluid to the blood stream.

The ECG changes depend on the level of rise of potassium.

- Mild rise (5.5 - 6.5 mEq/L): QT interval is shortened, T waves are tall and peaked, and ST segment is inverted.
- Moderate increase (6.5 - 8 mEq/L): shows decreased or absent P waves, PR interval is prolonged, QRS interval is widened, R wave is amplified, and T wave is peaked.
- High increase (>8.0 mEq/L): shows absence of P wave, QS is widened, and conduction blockade is seen. In later stages, patterns of ventricular fibrillation or even asystole may emerge.

Management:

The management depends on the severity of hyperkalemia.

Mild to moderate hyperkalemia:

- If the patients do not have any cardiac symptoms and ECG abnormalities, aggressive management is not required.
- Diuretics may be administered to promote excretion of potassium.
- If necessary, cation exchange resins may be administered. These are small beads which allow exchange of potassium for hydrogen ions.

Severe hyperkalemia:

This must be managed aggressively. There are three arms of treatment for severe hyperkalemia:

- Prevention of cardiac toxicity:
- Intravenous calcium administration protects the heart muscle from arrhythmias and arrest.
- Increasing potassium uptake by cells:
- This is done by administering insulin and glucose infusion. Insulin tends to increase potassium uptake by the cells.
- Increasing potassium excretion by the kidneys:

Chapter 8: Potassium: Homeostasis and Imbalances

- Loop diuretics may be administered. This must be done only if the renal function is normal.
- In emergency cases, potassium may be removed by dialysis.
- Other measures:
- If the patient is in metabolic acidosis, it must be corrected with bicarbonates or beta adrenergic agonists.

SUMMARY:

- Potassium is the predominant intracellular cation, and it plays an important role in intracellular osmolarity, maintaining resting membrane potential, and acid-base balance.
- Potassium levels in the serum are largely controlled by aldosterone.
- Hypokalemia may be mild, moderate, or severe, and may be caused by decreased intake, excessive excretion, or intracellular shift of potassium.
- It is managed by correcting the cause, and replenishing the body stores with intravenous potassium, at 10 mEq per hour.
- Hyperkalemia occurs mostly with renal dysfunction, or massive cell lysis.
- It can have serious cardiac effects, and calcium gluconate must be given to protect the heart. Excess potassium can be removed with diuretics, or by dialysis.

REFERENCES:

1. Marino PL. Potassium. The ICU Book. Baltimore: Williams & Wilkins; 1998

2. Palmer BF. Regulation of Potassium Homeostasis. Clinical Journal of the American Society of Nephrology : CJASN. 2015;10(6):1050-1060.

3. Greenlee M, Wingo CS, McDonough AA, Youn JH, Kone BC. Narrative review: evolving concepts in potassium homeostasis and hypokalemia. Ann Intern Med. May 2009. 150:619-625

4. Chew HC, Lim SH. Electrocardiographical case. A tale of tall T's. Hyperkalaemia. Singapore Med J. 2005 Aug. 46(8):429-32

5. Gumz ML, Rabinowitz L, Wingo CS. An Integrated View of Potassium Homeostasis. N Engl J Med. 2015 Jul 2. 373 (1):60-72

Chapter 8: Potassium: Homeostasis and Imbalances

EXERCISES:

1. Where are most of our potassium reserves stored?

    a. extracellular fluids

    b. intracellular fluids

    c. plasma

    d. bones

2. Which of the following is not a function of potassium?

    a. Maintaining resting membrane potential

    b. Contributes to acid-base balance

    c. Aids in blood clotting

    d. Main contributor to intracellular osmolarity

3. What is the normal concentration of potassium in the blood?

    a. 0.1-0.3 millimoles per liter

    b. 3.0-7.2 millimoles per liter

    c. 3.5-5.2 millimoles per liter

    d. 5.5-6.0 millimoles per liter

4. What is an example of the signs and symptoms of hypokalemia?

    a. Muscle pain

    b. Weakened bones

    c. Hallucinations

    d. Palpitations

Fluids and Electrolytes

5. Intracellular shift of potassium may be seen in all except:
    a. Refeeding syndrome
    b. Starvation
    c. Renal failure
    d. Insulin therapy

6. What infusion is given to drive potassium into the cells?
    a. Insulin
    b. Glucose
    c. Both the above
    d. Neither of the above

7. In which part of the kidney is potassium reabsorbed?
    a. Proximal convoluted tubule
    b. Descending loop of Henle
    c. Ascending loop of Henle
    d. Distal convoluted tubule

8. Hyperkalemia is caused by all the following except:
    a. Excess dietary intake of salt substitutes
    b. Addison's disease
    c. Infusion of packed cells
    d. Drugs like diuretics

Chapter 9:

Calcium: Homeostasis and Imbalances

Calcium is the most abundant cation present in the body as it forms most of the body's skeletal framework in the form of calcium phosphate. However, it has a limited concentration in the intracellular and extracellular fluids. Calcium has multiple uses in the body. Most people know of its importance maintaining the integrity and strength of bones. It also helps in activation of several proteins, including clotting factors in the blood. It is important for the proper functioning of nerves. Calcium levels in the extracellular fluid determine the sensitivity of voltage gated sodium ion channels. Low levels can cause hyperexcitability and high levels can cause depression of nerves. It serves as a secondary messenger for cell signaling. It also plays an important role in muscle contraction.

REGULATION OF CALCIUM:

Calcium enters the body through dietary intake and gets excreted through urine and feces. A large portion of calcium that enters the body is taken up by the bones, the rest exists in the body fluids.

The normal value of calcium in the extracellular fluid is around 2.2 – 2.6 mmol/L. This exists in two forms. Approximately half of this exists in the free or ionized form (1.1 – 1.3 mmol/L). This form is biologically active. The remainder, which is biologically inactive, exists in a bound form to various proteins and sulphates, the most common protein being albumin.

Dietary calcium is absorbed from the small intestine under the influence of Vitamin D (calcitriol). Excess calcium in the body is excreted by the kidneys, under the influence of the parathyroid hormone. The bones of the body act as reservoirs of calcium ions in the body, and can take up calcium ions from the blood or secrete them in to the blood as needed.

The level of ionized calcium in the plasma has to be set within extremely narrow limits in order to avoid excitation or depression of nerves. This is achieved by the action of two hormones – parathyroid hormone (PTH) which is secreted by the parathyroid gland, and calcitonin, which is secreted by the parafollicular cells of the thyroid gland.

When the blood levels of ionized calcium are higher than usual, the following actions take place:

- Calcitonin secretion increases, which in turn causes the bones to take up calcium from the blood into storage.
- PTH secretion is suppressed, which in turn has the following effects:
    - Inhibition of the removal of calcium from the bones.
    - Increased excretion of calcium in urine.
    - Decreases the excretion of phosphate in urine. This increases blood levels of phosphate, which combine with calcium, thereby decreasing the concentration of ionized calcium.
    - Also inhibits the activation of Vitamin D (Chlolecalciferol to calcitriol), which has two actions – it prevents further intestinal absorption of calcium, and

Chapter 9: Calcium: Homeostasis and Imbalances

inhibits bone osteoclasts from releasing calcium into the blood.

If the level of ionized calcium in the blood falls below normal limits, the following actions take place:

- PTH secretion is stimulated, which has the following effects:
- Bone is stimulated to release calcium into the blood
- Decreased excretion of calcium, and increased excretion of phosphate in urine.
- Vitamin D is activated, leading to more intestinal absorption of calcium
- Calcitonin secretion is suppressed, which prevents the bone from taking up calcium.

IMBALANCES:

HYPOCALCEMIA:

Hypocalcemia occurs when the total blood calcium level falls below 2 mmol/L. This may be a fall in ionized calcium alone ('true' hypocalcemia) or may reflect a fall in the unionized portion of calcium.

Causes:
- Deficiency of parathyroid hormone: Primary or secondary hypoparathyroidism, pseudoparathyroidism.
- Vitamin D deficiency: Rickets, osteomalacia
- Hypoalbuminemia: malnutrition, sepsis, burns etc.
- Disorders of other electrolytes such as hyperphosphatemia or hypomagnesemia
- Systemic diseases: Hepatic disease (impairs albumin), renal disease (impairs Vitamin D activation) etc.
- Drugs: Antiepileptic drugs, foscarnet etc.
- Following gastric surgery: impairs calcium absorption

Fluids and Electrolytes

Signs and Symptoms:
- Earliest symptoms are numbness and tingling sensations
- Muscle spasms can occur. Classic manifestation is carpopedal spasm, referred to as tetany.
- Laryngospasm and bronchospasm may occur, leading to wheezing and breathlessness
- Irritability and depression
- Later stages – seizures
- Signs of hypocalcemia:
    - Chvostek sign: Tapping the skin over the facial nerve causes ipsilateral contraction of the facial nerve muscles.
    - Trousseau's sign: A blood pressure cuff is used on the upper arm and inflated 20mmHg above systolic pressure. There is flexion at the wrist and metacarpo-phalangeal joints.

Management:
- In cases of mild hypocalcemia, oral calcium supplementation is all that is required. The cause must be identified and corrected. Serum levels of parathyroid hormone and Vitamin D must be analyzed and corrected.
- Severe hypocalcemia, which presents with tetany, seizures or arrhythmias, must be corrected with intravenous calcium. Calcium gluconate or calcium chloride is infused over 5 to 10 minutes.
- The serum levels must be monitored regularly to avoid overcorrection. In cases where the cause is hypoalbuminemia, ionized calcium must be measured separately.
- In chronic cases, regular calcium supplementation must be taken, along with supplements of Vitamin D and adequate exposure to sunlight.

Chapter 9: Calcium: Homeostasis and Imbalances

HYPERCALCEMIA:

This occurs when the blood levels of calcium rise beyond 2.5 mmol/L. This can be due to either impaired excretion of calcium by the kidneys, or due to too much calcium entering the blood from the bone reservoirs. Hypercalcemia may be categorized as mild (2.5 - 3.0 mmol/L), moderate (3.0 - 3.5 mmol/L), and severe, also called hypercalcemic crisis (more than 3.5 mmol/L)

Causes:

- Malignancies: This is seen in bone metastases, that cause increased osteoclastic activity, and tumors that secrete parathyroid hormone related peptide.
- Hyperparathyroidism: Tumors or hyperplasia of the parathyroid gland
- Vitamin D toxicity.
- Other conditions that cause high turnover of bone.

Signs and Symptoms:

Hypercalcemia affects several organs of the body; clinical manifestations are produced accordingly. This is popularly described as 'stones, bones, abdominal moans, and psychic groans'. These are described in detail below:

- Renal manifestations: Excess calcium can lead to the formation of renal calculi. This can cause dehydration, pain in the lower abdomen, and even renal failure.
- Gastrointestinal manifestations: Nausea, vomiting, constipation, and gastric ulcers.
- Central nervous system manifestations: Lethargy, confusion, depressed consciousness.
- Cardiac manifestations: Since calcium is believed to have a positive inotropic effect, excess calcium in the blood can cause arrhythmias.

Management:

Management of hypercalcemia depends on the cause and the severity.

Medical management:

- Isotonic sodium chloride can flush out excess calcium temporarily and is good for short-term management
- Loop diuretics – increase calcium excretion
- Bisphosphonate drugs – prevent osteoclastic activity
- Dialysis: both peritoneal dialysis and hemodialysis are effective in immediately lowering serum calcium levels.

Surgical management:

- Tumors responsible for raised calcium levels must be resected.
- Primary hyperparathyroidism requires surgical removal of the parathyroid glands.

SUMMARY:

- Calcium is an important cation that makes up the body's structural framework. It also acts as a secondary messenger and is essential for muscle contraction.
- Calcium homeostasis is regulated by the interaction between Vitamin D, parathyroid hormone, and calcitonin
- Hypocalcemia is mainly caused by Vitamin D deficiency, or impaired PTH secretion or function.
- It is manifested by numbness, muscle spasms such as carpopedal spasm, Chvostek's sign and Trousseau's sign
- It is treated by oral or intravenous calcium supplementation.

Chapter 9: Calcium: Homeostasis and Imbalances

- Hypercalcemia is due to malignancies or Vitamin D toxicity. It manifests with renal stones, bone pain, and gastric pain.
- Treatment is by diuretics, dialysis, and bisphosphonate drugs.

REFERNCES:

1. Mundy GR, Guise TA. Hormonal control of calcium homeostasis. Clin Chem. 1999 Aug. 45(8 Pt 2):1347-52

2. Cooper MS, Gittoes NJ. Diagnosis and management of hypocalcaemia. BMJ. 2008 Jun 7. 336(7656):1298-302

3. Goyal A, Bhimji SS. Hypocalcemia. [Updated 2017 Apr 25]. In: StatPearls [Internet]. Treasure Island (FL): StatPearls Publishing; 2017 Jun-. Available from: https://www.ncbi.nlm.nih.gov/books/NBK430912

4. Goldner W. Cancer-Related Hypercalcemia. J Oncol Pract. 2016 May. 12 (5):426-32.

5. Makras P, Papapoulos SE. Medical treatment of hypercalcaemia. Hormones (Athens). 2009 Apr-Jun. 8(2):83-95

6. Turner JJO. Hypercalcaemia - presentation and management . Clin Med (Lond). 2017 Jun. 17 (3):270-273

Chapter 9: Calcium: Homeostasis and Imbalances

EXERCISES:

1. Where are most of our calcium reserves stored?
    a. Blood
    b. Muscle tissues
    c. Bones
    d. Intracellularly

2. All the following are functions of calcium, except:
    a. Bone metabolism
    b. Muscle contraction
    c. Clotting of blood
    d. Regulation of body temperature

3. What is the normal concentration of calcium in the blood?
    a. 0.2-0.9 millimoles per liter
    b. 2.2-2.6 millimoles per liter
    c. 6.0-10.0 millimoles per liter
    d. 10.0 -12.0 millimoles per liter

4. What is the condition where there is an increased levels of calcium in the blood?
    a. Hypocalcemia
    b. Hypercalcemia
    c. Hypercalciuria
    d. Hypocalciuria

Fluids and Electrolytes

5. What is an example of the signs and symptoms of hypocalcemia?
    a. Loss of appetite
    b. Muscle spasm
    c. Hair loss
    d. Increase in body temperature

6. Chvostek sign is used to stimulate which of the following nerves?
    a. Facial nerve
    b. Optic nerve
    c. Trigeminal nerve
    d. Glossopharyngeal nerve

7. Which is not a treatment modality for hypercalcemia?
    a. Hypotonic saline
    b. Bisphosphonates
    c. Loop diuretics
    d. Dialysis

8. Which organ system is not affected by hypercalcemia?
    a. Renal system
    b. Cardiovascular system
    c. Respiratory system
    d. Gastrointestinal system

Chapter 10:

Magnesium: Homeostasis and Imbalances

Magnesium is the second most abundant intracellular cation, after potassium. It acts as a cofactor for several enzymes, and therefore plays an important role in cellular processes such as transcription of DNA, synthesis of proteins, and energy metabolism. It is necessary for bone formation, as part of hydroxyapatite, and also for regulation of muscle contraction.

REGULATION OF MAGNESIUM BALANCE:

The normal level of magnesium in the blood is 0.7 -1.1 mmol/L. Magnesium balance is maintained in the body by coordination of function between the intestine, bones, and kidneys.

The source of magnesium for the body is through dietary intake, which is usually around 30 mg per day. Magnesium absorption is carried out by the cells of the small intestine. Most of the absorption is passive, through transcellular transport. If the dietary intake increases, active absorption takes place, though paracellular transport. Proton pump inhibitors can decrease absorption of magnesium from the small intestine.

About 50% of body magnesium is stored in the bones. It is also stored in muscle and soft tissue. When the blood levels are lowered, magnesium is released from these sources.

The kidney filters almost 2000 mg of magnesium per day. However, most of this is reabsorbed. 10-30% of magnesium gets reabsorbed in the proximal convoluted tubule, 40-70% of magnesium gets reabsorbed in the thick ascending loop of Henle, and 5 to 10% gets reabsorbed in the distal convoluted tubule.

There are several factors which can affect renal reabsorption of magnesium.

Increased reabsorption of magnesium:

- Hormones – parathyroid hormone, calcitonin, glucagon, insulin, aldosterone
- Metabolic alkalosis
- Epidermal growth factor

Decreased reabsorption of magnesium:

- Increased blood levels of calcium and magnesium
- Prostaglandin E2
- Metabolic acidosis
- Drugs – diuretics, antibiotics, immunosupressants

IMBALANCES:

HYPOMAGNESEMIA

This occurs when the plasma magnesium level falls below 0.7 mmol/L. Hypomagnesemia is an important condition that can cause disturbances in virtually every organ system in the body. It has been associated with several systemic diseases and can cause other electrolyte disturbances as well.

Chapter 10: Magnesium: Homeostasis and Imbalances

Causes:
- Reduced dietary intake of magnesium: starvation, intravenous total parenteral nutrition, or alcoholism
- Reduced absorption of magnesium by the small intestine: drugs such as proton pump inhibitors
- Shift of magnesium from the extracellular to intracellular space: refeeding syndromes, diabetic ketoacidosis, and alcohol withdrawal syndrome.
- Increased excretion: due to decreased reabsorption of magnesium from the thick ascending loop of Henle and distal convoluted tubule. This can occur in several conditions, described in the previous section on magnesium regulation, and also in autosomal recessive disorders such as Gilteman's syndrome.

Signs and symptoms:
- Weakness, fatigue, lethargy
- Numbness
- Hyper-reflexes
- Muscle cramps and tetany
- Convulsions
- Nausea, vomiting
- Arrythmias
- ECG changes are seen, which include T wave changes, prolonged QT interval, and abnormal patterns of ventricular tachycardia, ventricular fibrillation, or torsades de pointes.

Consequences:
Magnesium deficiency has been implicated in several systemic diseases, such as:

- Hypertension
- Diabetes
- Coronary artery disease

- Osteoporosis
- Renal stones and nephrolithiasis

Sudden magnesium deficit can have fatal consequences due to ventricular arrhythmias and coronary artery vasospasm.

It is also associated with other electrolyte imbalances such as:

- Hypokalemia: due to increased excretion of potassium
- Hypocalcemia: this is secondary due to impaired PTH release.

Management:
- Mild cases can be corrected by increasing the dietary intake alone.
- Oral replenishment for chronic cases. This includes oral intake of magnesium oxide, or magnesium chloride, at a dose of 60-80 mg (3.5 mmol), at least six to eight times a day.
- Acute and life threatening cases must undergo intravenous replacement. 50 mEq of magnesium can be given over 8 to 24 hours, until the blood level reaches at least 0.4 mmol/L.
- It is important to monitor serum levels of potassium and calcium, and correct these if required.

HYPERMAGNESEMIA:

Hypermagnesemia occurs when the serum magnesium levels cross 1.2 mmol/L. However, symptoms occur only if the serum level crosses 2 mmol/L, and the manifestations increase in severity with increasing levels.

Causes:
- Increased ingestion of magnesium: Due to over-the-counter medications such as Epsom salts, antacids, and laxatives that contain high amounts of magnesium.

Chapter 10: Magnesium: Homeostasis and Imbalances

- Excess iatrogenic administration: During management of conditions associated with hypomagnesemia.
- Renal failure: prevents normal excretion of magnesium
- Tissue breakdown that releases intracellular magnesium ions into the blood: sepsis, burns

Signs and symptoms
- Neuromuscular manifestations:
- Absence or depressed deep tendon reflexes
- Muscle weakness and paralysis (above 5 mmol/L)
- Facial paresthesia
- Respiratory muscle weakness- apnea
- Decreased cardiac conduction – bradycardia, heart block, and even cardiac arrest (above 7 mmol/L)
- Bleeding: prevents platelet activation and thrombin generation, thereby prolongs bleeding

Management:
- To prevent cardiac toxicity: Intravenous calcium gluconate is given.
- To remove excess magnesium from body: Diuretics such as furosemide are used, or dialysis may be done.
- Promote intracellular uptake: A combination of glucose and insulin infusion will help to drive magnesium ions into the cells.

SUMMARY:

- Magnesium is an important intracellular cation that is a co-factor for enzymes, and is involved in DNA transcription, protein, and energy metabolism.
- Homeostasis of magnesium is regulated by several hormones including parathyroid hormone, calcitonin, insulin, aldosterone, and epidermal growth factor.
- Hypomgnesemia affects every organ system in the body and plays a role in pathogenesis of several systemic diseases.
- It may be corrected by dietary intake, oral supplements, or intravenous supplements.
- Hypermagnesemia is mainly caused by antacid overdose, especially Epsom salt.
- Therapy includes protecting the heart with calcium infusion, using diuretics and dialysis, and driving magnesium intracellularly with insulin-glucose.

REFERENCES:

1. Glasdam SM, Glasdam S, Peters GH. The Importance of Magnesium in the Human Body: A Systematic Literature Review. Adv Clin Chem. 2016. 73:169-93

2. Blaine, J., Chonchol, M., & Levi, M. (2014). Renal Control of Calcium, Phosphate, and Magnesium Homeostasis. Clinical Journal Of The American Society Of Nephrology, 10(7), 1257-1272.

3. Weber, C., & Santiago, R. (1989). Hypermagnesemia. Chest, 95(1), 56-59. De Baaij, J. H. F., Hoenderop, J. G. J., & Bindels, R. J. M. (2012). Regulation of magnesium balance: lessons learned from human genetic disease. Clinical Kidney Journal, 5(Suppl 1), i15–i24.

4. Martin KJ, González EA, Slatopolsky E. Clinical consequences and management of hypomagnesemia. J Am Soc Nephrol. 2009 Nov. 20(11):2291-5

5. Bringhurst FR, Demay MB, Krane SM, et al. Bone and mineral metabolism in health and disease/hypermagnesemia. Kasper DL, et al, eds. Harrison's Principles of Internal Medicine. 16th ed. New York, NY: McGraw-Hill; 2005. 2245

EXERCISES:

1. What is the normal level of magnesium in the blood?
    a. 0.7 - 1.1 mmol/l
    b. 1.7 - 2.1 mmol/l
    c. 2.7 - 3.1 mmol/l
    d. 3.7 - 4.1 mmol/l

2. Magnesium is stored in all except which one of the following body tissues?
    a. Bone
    b. Muscle
    c. Soft tissue
    d. Blood

3. In which part of the kidney does maximum magnesium reabsorption take place?
    a. Proximal convoluted tubule
    b. Loop of Henle
    c. Distal convoluted tubule
    d. Collecting duct

4. Which of the following hormones is not involved in the regulation of magnesium?
    a. Aldosterone
    b. Insulin
    c. Adrenaline
    d. Calcitonin

Chapter 10: Magnesium: Homeostasis and Imbalances

5. Which of the following factors causes decreased reabsorption of magnesium?

    a. Parathyroid hormone

    b. Epidermal growth factor

    c. Metabolic alkalosis

    d. Prostaglandin E2

6. What is the normal intravenous dose of magnesium for correction of hypomagnesemia?

    a. 20 mEq

    b. 30 mEq

    c. 40 mEq

    d. 50 mEq

7. All the following are systemic diseases that are associated with low magnesium levels, except:

    a. Osteoporosis

    b. Diabetes mellitus

    c. Rheumatoid arthritis

    d. Hypertension

8. Which of the following electrolytes can show abnormalities with hypomgnesemia?

    a. Sodium and potassium

    b. Potassium and calcium

    c. Calcium and sodium

    d. Calcium and phosphorous

Fluids and Electrolytes

9. The following are signs and symptoms associated with hypermagnesemia, except:
    a. Hyper reflexes
    b. Facial paresthesia
    c. Bradycardia
    d. Bleeding

10. The management of hypermagnesemia involves all of the following, except:
    a. Calcium gluconate therapy
    b. Diuretics
    c. Intravenous magnesium
    d. Insulin- glucose infusion

Chapter 11:

Phosphorous: Homeostatis and Imbalances

Phosphorous is a prominent intracellular anion and has several important functions. It contributes to the structural integrity of cells as a component of the cell membrane and nucleic acids. It forms the end product and reservoir of energy metabolism (Adenosine triphosphate or ATP), and it activates several enzymes by phosphorylating them. It plays an important role in maintaining acid-base balance of the body, and contributes to bone mineralization.

REGULATION OF PHOSPHOROUS LEVELS:

The normal blood level of phosphorus is 0.8 to 1.45 mmol/L. Phosphorous is acquitted through dietary sources, which should ideally be around 1000mg to 1500 mg per day. Absorption primarily takes place in the jejunum area of the small intestine. Some amount of phosphorous also gets secreted into the intestine by the bile and pancreatic juice, which is then absorbed from the intestine. Phosphorous absorption takes place both passively, through diffusion, and actively, through sodium dependent transporters (NaPiIIa, NaPiIIb, and NaPiIIc). Vitamin D (calcitriol) plays an important role in enhancing phosphorous absorption from the small intestine.

In the body, phosphorous is stored primarily in the bones and teeth, and to a certain extent in soft tissues. There is a need-based movement of phosphorous between the bone cells and blood. Whenever bone mineralization takes place, phosphorous is taken up from the blood stream, and whenever there is resorption of bone, phosphorous is released into the bloodstream.

Excretion of phosphorous takes place primarily through the kidneys. This is completely filtered at the glomerular level, and is reabsorbed in the proximal convoluted tubule. The amount of phosphorous reabsorbed in the proximal tubule is negatively dependent on parathyroid hormone. High PTH expression decreases reabsorption and low levels stimulate reabsorption.

Another group of regulators, called phosphatonins, are believed to regulate the metabolism of phosphorous. Of these, the phosphatonin which has been scrutinized closely is the fibroblast growth factor-23 (FGF-23), whose expression is believed to be increased with increased dietary intake of Vitamin D. FGF-23, like PTH, also decreases renal reabsorption of phosphorous, and promotes excretion. It is believed to suppress calcitriol activity, and, to a lesser extent, PTH.

FGF-23, calcitriol, and PTH are all interrelated as follows:

- Calcitriol stimulates expression of FGF-23
- FGF-23 suppresses both calcitriol, and PTH.
- PTH stimulates calcitriol.

HYPOPHOSPHATEMIA

Hypophosphatemia is a condition where the phosphate levels in the blood fall below 0.8 mmol/L.

Chapter 11: Phosphorous: Homeostatis and Imbalances

Causes:

- Inadequate intestinal absorpttion: Although uncommon, it may occur in Vitamin D deficiency, poor diet, and excessive intake of antacids.

- Increased excretion: Hyperparathyroidism, Phosphate wasting syndromes such as Fanconi syndrome.

- Shift from the extracellular to intracellular space: Refeeding syndrome, diabetic ketoacidosis, respiratory alkalosis etc.

Signs and symptoms:

These are extremely non-specific. Mild cases are not symptomatic. Symptoms may be seen in severe cases or long standing cases. These include:

- Weakness
- Bone pain
- Rhabdomyolysis: muscle breakdown and pain.
- Altered mental status

If the phosphate imbalance is not managed well and blood phosphate levels continue to go down, the patient may experience some psychological effects. Some patients may experience increased irritability and confusion. Some patients also show slurred speech.

If left unmanaged, the patient may have chest pains and cardiac dysrhythmia. It will be followed by increased breathing rate. In worst cases, the condition may lead to seizures and coma.

Management:

The best way to deal with low phosphate levels is to reintroduce the electrolyte back in the system. In hospitals and other medical facilities, this can be done with intravenous potassium phosphate. This is done to people whose phosphate levels are severely low, or in life threatening situations, such as hypocalcemia with tetany, hyperkalemia, and metabolic acidosis.

For those who have mild hypophosphatemia, oral phosphate supplements may also be used. This is also used in places where intravenous methods are not available.

Supportive therapy, such as Vitamin D supplementation, and concomitant correction of calcium and potassium imbalances, is essential to treatment of hypophosphatemia.

As with other electrolyte imbalances, hypophosphatemia will continue unless the underlying condition is addressed.

The ideal modality of management is outlined in the table below:

TABLE 1. MANAGEMENT OF HYPOPHOSPHATEMIA		
SEVERITY OF HYPOPHOSPHATEMIA	BLOOD LEVELS	MODALITY OF MANAGEMENT
Severe – critically ill patients, and	< 0.3 mmol/L	Intravenous phosphate infusion, at a rate of 0.08 to 0.16 mmol/kg, over 2 – 6 hours
Moderate – in ventilated patients	0.3 – 0.8 mmol/L	
Moderate – in non-ventilated patients, and	0.3 – 0.8 mmol/L	Oral phosphate supplementation alone
Mild	0.8 mmol/L	

HYPERPHOSPHATEMIA

Hyperphosphatemia is said to occur when the serum phosphate concentration exceeds 1.46 mmol/L. It is usually, but not always associated with hypercalcemia.

Causes:
- Excessive intake: either through diet or intravenous therapy
- Shift from intracellular to extracellular space: This occurs when cells are destroyed.

Chapter 11: Phosphorous: Homeostatis and Imbalances

- - Crush injuries
  - Rhabdomyolysis
  - Hemolytic anemia
  - Drugs – Chemotherapeutic agents, which are cytotoxic etc.
- Metabolic or respiratory acidosis: tends to shift phosphorous between cells
- Decreased excretion of phosphorous:
  - Hypoparathyroidism or resistance to PTH in Vitamin D toxicity, or magnesium imbalance.
  - Renal failure or insuffieciency

Signs and symptoms

Initially, hyperphosphatemia is asymptomatic, or presents with vague, nonspecific symptoms.

- Fatigue, weakness
- Nausea, vomiting, anorexia
- Shortness of breath
- Disturbances in sleep and insomnia.

Later on, signs of hypocalcemia may manifest, including the following:

- Trousseau's sign and Chvostek's sign
- Carpopedal spasms
- Hyper-reflexes
- Seizures and convulsions.

Management:

The cause must be identified in order to administer correct treatment.

Intake of dietary phosphate must be restricted.

Patients can take medications that bind with phosphorous and thereby reduce their blood levels. Phosphate binders are compounds that limit the absorption of phosphate from food, and need to be taken with each meal. As it travels through the gastrointestinal tract, the active agent in the drug binds with the protein. They are excreted with the food eaten because the bound phosphate is insoluble. There are several types of phosphate binders. Lanthanum carbonate and Sevelamer compounds are preferred. Calcium and aluminium based binders are also available, but are generally avoided as they can carry their own side effects.

If the cause of hyperphosphatemia is tumor lysis, and if the renal function is intact, saline diuresis can be done to wash off excess phosphate from the body.

In cases of renal dysfunction, hemodialysis may be used for removal of excess phosphorous from the body.

SUMMARY:

- Phosphorous is an intracellular cation, that is important for structural integrity of the cell and cell components.
- Homeostasis of phosphorous is achieved by regulation between Vitamin D, parathyroid hormone, and phosphatonins like FGF-23
- Hypophosphatemia is usually associated with hypercalcemia and hyperphosphatemia is associated with hypocalcemia.
- Hypophosphatemia occurs with Vitamin D deficiency and hyperparathyroidism. It can be managed by oral supplementation, intravenous being reserved for severe cases.
- Hyperphosphatemia occurs when parathyroid hormone secretion is reduced or resistance develops, or during cell lysis
- Depending on the cause, it is managed with phosphate binder, saline diuresis or dialysis.

Chapter 11: Phosphorous: Homeostatis and Imbalances

REFERENCES:

1. Penido, M. G. M. G., & Alon, U. S. (2012). Phosphate homeostasis and its role in bone health. Pediatric Nephrology (Berlin, Germany), 27(11), 2039-2048.

2. Bergwitz, C., & Jüppner, H. (2010). Regulation of Phosphate Homeostasis by PTH, Vitamin D, and FGF23. Annual Review of Medicine, 61, 91-104.

3. Liamis G, Milionis HJ, Elisaf M. Medication-induced hypophosphatemia: a review. QJM. 2010 Jul. 103(7):449-5

4. Block GA, Wheeler DC, Persky MS, et al. Effects of phosphate binders in moderate CKD. J Am Soc Nephrol. 2012 Aug. 23(8):1407-15

EXERCISES:

1. What is the normal range of phosphate in the blood?

    a. about 0.8 to 1.4 mmol/L

    b. about 0.2 to 0.8 mmol/L

    c. about 0.3 to 0.5 mmol/L

    d. about 0.6 to 2.5 mmol/L

2. Absorption of phosphate takes place in:

    a. Duodenum

    b. Jejunum

    c. Ileum

    d. Colon

3. Which of the following is an example of Phosphatonin?

    a. Calcitriol

    b. Epidermal growth factor 23

    c. Fibroblast growth factor 23

    d. Parathyroid hormone

4. Which of the following statements is true?

    a. Calcitriol depresses expression of FGF-23

    b. FGF-23 suppresses both calcitriol, and PTH.

    c. FGF-23 stimulates both calcitriol, and PTH.

    d. PTH depresses calcitriol

Chapter 11: Phosphorous: Homeostatis and Imbalances

5. Which serum level of phosphate indicates severe hypophosphatemia?

    a. Less than 0.8 mmol/L

    b. Less than 0.7 mmol/L

    c. Less than 0.5 mmol/L

    d. Less than 0.3 mmol/L

6. Which one of the following is a sign or symptom of hypophosphatemia?

    a. Itchiness of the skin

    b. Bone pain

    c. Shortness of breath

    d. Bloating of extremities

7. How is mild Hypophosphatemia treated?

    a. Phosphate supplements

    b. Phosphate binders

    c. Taking diuretics

    d. Dialysis

8. Intravenous supplementation of phosphorus is indicated in which one of the following situations?

    a. Moderate hypophosphatemia in non-ventilated patients

    b. Moderate hypophosphatemia in ventilated patients

    c. Mild hypophosphatemia

    d. Intravenous supplementation is never indicated

Fluids and Electrolytes

9. In which of the following situations can hyperphosphatemia occur?
    a. Metabolic acidosis
    b. Respiratory acidosis
    c. Both of the above
    d. Neither of the above

10. Which of the following is preferred as phosphate binder?
    a. Lanthanum compounds
    b. Calcium compounds
    c. Aluminum compounds
    d. Strontium compounds

Chapter 12:

Imbalances in Common Medical Conditions

Now that an overview has been provided into fluid and electrolyte imbalances, it would have been noted that the cause of most of these imbalances is similar. Generally, dysfunction of any one of the major organ systems, or insult to the body as a whole would bring about changes in the normal homeostatic mechanisms, leading to fluid, electrolyte, and acid-base disorders. These disorders would occur in combination, rather than in isolation. A few of the common medical conditions in which these would present are described in this chapter.

HYPERTHERMIA:

Hyperthermia, or heat stroke, is a condition where the body heats up because thermoregulation fails. This can occur in the presence of excessive environmental heat, excessive exertion, or a combination of both.

Pathophysiology of imbalances:
- Heat causes the body to lose more fluid, both through sweat and insensible losses. Therefore, the patient develops dehydration.

- As fluid loss occurs, the concentration of electrolytes relatively increases. This can lead to hypernatremia.
- Hypokalemia is seen in early heat stroke due to fluid and electrolyte redistribution. However, in later phases, as muscle damage increases, potassium is released from damaged cells leading to hyperkalemia.
- Hypocalcemia and hypomagnesemia are observed, secondary to other electrolyte disturbances.
- Lactic acidosis may occur because of poor aerobic metabolism and accumulation of lactates.

Management:
- Immediate cooling of the body is the mainstay of treatment, and can halt or reverse the disease process. This is accomplished by ice water immersion, or evaporative cooling therapy.
- Infusion of dextrose with sodium bicarbonate can provide fluid resuscitation, drive potassium intracellularly, and counter acidosis.
- Calcium must be used only if cardiac signs develop.

CARDIAC FAILURE:

Cardiac failure occurs when the heart muscle is unable to pump sufficient blood to meet the requirements of the entire body. The various neurohumoral adaptive mechanisms that take place in heart failure, as well as the various drugs used, can contribute to electrolyte and acid-base disorders.

Pathophysiology of imbalances:
- Hyponatremia: when the cardiac output decreases, the renal blood flow and hence glomerular filtrate also decreases, leading to water retention. Apart from this, there is activation of renin-angiotensin mechanism as well as the ADH hor-

Chapter 12: Imbalances in Common Medical Conditions

mone, which leads to more retained water, thereby diluting the sodium content present.

- Hypokalemia: this can occur secondary to renin-angiotensin activation, and can also be a consequence of diuretic therapy. Hypokalemia is a strong predictor of mortality rate as by itself, it can cause cardiac arrhythmias in these patients.

- Hypomagnesemia: although this is commonly observed in patients with heart failure, its pathophysiology is less well understood. It is believed to be due to reduced dietary intake, renal losses due to diuretics, and due to redistribution of other electrolytes.

- Hypocalcemia and hypophosphatemia: These do not always occur, but may be seen sometimes secondary to other electrolyte disturbances.

- Acid-base disturbances: Usually these patients can present with metabolic alkalosis, sometimes in combination with respiratory alkalosis. This is mainly due to loss of hydrogen ions in the urine, and driving of hydrogen ions intracellularly. Hypoxia, due to reduced circulating blood, is also a contributing factor.

Management:

- Since cardiac failure is a condition with high morbidity, focus is usually on treating the condition proper. Based on the severity, treatment for heart failure ranges from pharmacological therapy and revascularization procedures, to a total heart transplant.

- At every stage on therapy, however, the electrolytes and blood gases must be analyzed periodically, so that drug therapy may be modified, and any electrolyte imbalances may be corrected.

- It is especially important to keep potassium and magnesium levels within their normal ranges, as these electrolytes play an important role in heart health.

RESPIRATORY DISEASE:

Respiratory failure occurs when the lungs fail to facilitate exchange of carbon dioxide for oxygen. This leads to lack of oxygen (hypoxia) and excess of carbon dioxide (hypercapnea) in the body.

Pathophysiology of imbalances:
- Since the respiratory mechanism of acid-base homeostasis is suppressed, there is accumulation of carbon dioxide in the body. This results in respiratory acidosis due to raised PaCO2.
- Impaired lung function can eventually cause pulmonary edema and right heart dysfunction, which in turn affects the renin-angiotensin-aldosterone axis, and stimulates ADH release. The net effect is water retention, leading to hyponatremia.

Management:
- The mainstay of therapy is correcting the hypoxemia. The standard of treatment is extracorporeal membrane oxygenation, which reverses the hypoxia, and thereby the hypercapnea.
- Once these abnormalities are reversed, electrolytes must be monitored and corrected if required.
- Severe acidosis may require assisted mechanical ventilation to remove carbon dioxide from the body.

GASTROINTESTINAL FLUID LOSSES:

There are several instances where fluid may be lost from the gastrointestinal tract. While gastric fluid is acidic, the bile and pancreatic juice secretions are more alkaline. Thus the electrolyte and acid-base imbalance would depend on the type of fluid that is lost.

Chapter 12: Imbalances in Common Medical Conditions

Pathophysiology of imbalances:
- Vomiting and diarrhea leads to fluid loss, along with loss of sodium and potassium. This is further exacerbated by inadequate intake. Generally this manifests as hypokalemia and hypochloremia. Hypochloremia results in alkalosis.
- On the other hand in conditions like pancreatic fistulae, loss of pancreatic juice can cause losses of bicarbonate, along with sodium and potassium, leading to a net acidosis.
- Milk alkali syndrome: This is seen in peptic ulcer patients who have ingested large quantities of antacids. This is associated with hypercalcemia and mild alkalosis. Magnesium levels may also be raised.
- Steatorrhoea is a condition where there is impaired fat absorption. This also impairs the absorption of calcium and phosphorous, leading to hypocalcemia and hypophosphatemia.

Management:
Management depends on the cause of gastric fluid losses. Electrolyte monitoring according to the trends mentioned above is essential. Management must be tailored to meet the abnormality.

RENAL FAILURE:

Renal failure invariably causes electrolyte abnormalities across the entire spectrum, and can result in complex acid-base disturbances.

Pathophysiology of imbalances:
- Since there is failure of fluid excretion, there is fluid excess, leading to edema. Although sodium is also retained, the excessive fluid retention results in a net dilution and hyponatremia.
- Hyperkalemia: failure of secretion of potassium in the renal tubule.

- Kidneys are also responsible for activation of vitamin D, which in turn influence calcium levels. In renal failure, hypocalcemia and hyperphosphatemia is frequently encountered.
- In renal failure, magnesium fails to get excreted, leading to hypermagnesemia.
- Renal failure also prevents reabsorption of bicarbonate, resulting in metabolic acidosis.

Management:

The cause of renal failure must be determined and managed if possible.

- Diuretic drugs are frequently used to remove excess water and electrolytes. However, this must be done with caution as diuretic therapy itself can lead to electrolyte imbalances. Electrolyte monitoring is done frequently and drugs must be modified or changed as needed.
- Dialysis, both hemodialysis and peritoneal dialysis, are the mainstay of treatment and remove excess electrolytes.
- Oral bicarbonate supplementation is helpful in cases of chronic kidney failure. Calcium supplementation can be done orally or intravenously.

BURNS:

Pathophysiology of imbalances:

Below the large necrotic area of the burn, inflammatory mediators are released.

These cause increase in vascular permeability, causing plasma to shift out of vessels into the interstitial and intracellular space, causing cell edema. Both sodium and water are shifted intracelluarly.

Early phase:

Chapter 12: Imbalances in Common Medical Conditions

- Hyponatremia: due to extracellular sodium depletion
- Hyperkalemia: released during cell damage of burnt area.
- Acidosis : secondary to hyperkalemia

Second phase:

- Hypernatremia: mobilization of intracellular sodium, activation of rennin-angiotensin pathway
- Hypocalcemia: due to potassium losses and intracellular shift
- Hypomagnesemia and hypophosphatemia: due to loss of other electrolyes

Management:
- First the hyperkalemia is treated:
    - Calcium chloride infusion to counter cardiac effects of hyperkalemia
    - Drive potassium intracellularly with insulin and glucose infusion
- Sodium bicarbonate to correct acidosis if needed
- Use of diuretics/dialysis to excrete potassium
- If hypernatremia persists, it is corrected with hypotonic fluids
- All other electrolytes – calcium, phosphate, magnesium, must be observed every few hours and corrected if needed.

SUMMARY:

- Certain medical conditions cause imbalances of several electrolytes at once. As these symptoms are non-specific, it is important to be aware of this and monitor electrolytes periodically.

Fluids and Electrolytes

- Hyperthermia causes fluid loss due to heat, leading to hypernatremia and acidosis.
- Congestive cardiac failure causes edema and water retention, leading to hyponatremia and metabolic alkalosis
- Respiratory failure primarily causes respiratory acidosis
- Gastric fluid disturbances depend on the nature of fluid lost. Loss of gastric juice can cause alkalosis while loss of pancreatic fluid can cause acidosis
- Renal failure upsets almost all electrolytes as the kidneys play an important role in reabsorption of these. Hence all electrolytes must be monitored and managed periodically.
- Burns patients develop increased vascular permeability and fluid shifts interstitially and intracellularly. This causes hyponatremia and hyperkalemia. Acidosis can occur.

REFERENCES:

1. Noakes, T. D. (1998). Fluid and electrolyte disturbances in heat illness. International journal of sports medicine, 19(S 2), S146-S149.

2. Simon, H. B. (1993). Hyperthermia. New England Journal of Medicine, 329(7), 483-487.

3. Noakes, T. D. (1998). Fluid and electrolyte disturbances in heat illness. International journal of sports medicine, 19(S 2), S146-S149.

4. Moe, A. E. (1955). Electrolyte Balance in Gastrointestinal Disease. California Medicine, 83(5), 339–342.

5. Haberal, M., Sakallioglu Abali, A. E., & Karakayali, H. (2010). Fluid management in major burn injuries. Indian Journal of Plastic Surgery : Official Publication of the Association of Plastic Surgeons of India, 43(Suppl), S29–S36.

Fluids and Electrolytes

EXERCISES:

1. What is not an accepted mode of treatment for hyperthermia?

    a. Ice water immersion

    b. Evaporative therapy

    c. Warm saline infusion

    d. Bicarbonate therapy

2. Which of the following electrolyte abnormalities is not seen in cardiac failure?

    a. Hyponatremia

    b. Hypernatremia

    c. Hypokalemia

    d. Hypocalcemia

3. Respiratory failure causes which one of the following acid base disorders?

    a. Metabolic acidosis

    b. Metabolic alkalosis

    c. Respiratory acidosis

    d. Respiratory alkalosis

4. Which one of the following digestive secretions has an acidic pH?

    a. Saliva

    b. Gastric juice

    c. Bile

    d. Pancreatic juice

Chapter 12: Imbalances in Common Medical Conditions

5. Renal failure results in which one of the following acid base disorders?

    a. Metabolic acidosis

    b. Metabolic alkalosis

    c. Respiratory acidosis

    d. Respiratory alkalosis

6. Steathorrhoea would affect the absorption of which of the following electrolytes?

    a. Sodium

    b. Potassium

    c. Magnesium

    d. Calcium

7. In which of the following medical conditions are all the electrolytes invariably affected?

    a. Respiratory disease

    b. Hypertension

    c. Renal disease

    d. Nervous disease

8. In which phase of burns does hyponatremia occur?

    a. Early phase

    b. Late phase

    c. Both phases

    d. Hyponatremia does not occur

Fluids and Electrolytes

9. What is the reason for fluid shift in burns patients?
    a. Lymphatic block
    b. Increased vascular permeability
    c. Increased blood pressure
    d. None of the above

10. Irrespective of the medical condition, which electrolyte abnormality must be addressed first?
    a. Potassium
    b. Sodium
    c. Calcium
    d. Phosphorous

Conclusion

We hope this book was able to help you to learn about body fluids, electrolytes and the acid-base balance. By now, you must be familiar with the normal homeostatic mechanisms in place for regulation of fluids, electrolytes, and acid-base balance. You must also be familiar with recognizing the common disorders associated with disturbance in these homeostatic mechanisms. It must be remembered that correcting fluid and electrolyte imbalances in patients with these disorders will go a long way in reducing morbidity and mortality.

This book has outlined the basic management modalities for fluid and electrolyte disorders. However, it must be remembered that management of these disorders is often a challenging undertaking and must be approached with caution. Fluid resuscitation is the mainstay of therapy, but choice of fluid, and the rate and quantity of administration, is extremely critical. Rapid correction can lead to severe complications. Overcorrection can also lead to imbalances of other electrolytes in the system. Moreover, fluid administration is a dynamic process, and there is often high inter-individual variability. Therefore, in managing patients with these disorders, it is necessary not only to apply knowledge learned in this book, but also to take into account the patients' entire systemic condition. It is necessary to periodically test for levels of not only the electrolyte being corrected, but other electrolytes as well. As has been emphasized at various points, correction of the underlying cause is the best modality of treatment.

The next step is to keep on learning about how the imbalances discussed in this book affect your own specialization in the medical field. You should also come back to this book in the future to remind yourself of how fluid, electrolytes and the acid-base balance are interconnected. Fluids and electrolytes are often an extremely volatile subject, which needs constant refreshing in order to deliver better care. You can repeat the post-chapter exercises at periodic intervals to assess your memory.

Thank you and good luck!
David Andersson and the Medical Creations team

Answers to Exercises

CHAPTER I

1. b
2. c
3. b
4. a
5. d
6. c
7. c
8. a
9. c
10. c

CHAPTER II

1. c
2. a
3. c
4. c
5. b

6. d
7. c
8. b
9. b
10. b

CHAPTER III

1. c
2. c
3. b
4. d
5. b
6. c
7. a
8. d
9. c
10. a

CHAPTER IV

1. c
2. a
3. c
4. d
5. c
6. b
7. d
8. c

Answers to Exercises

9. a
10. b

CHAPTER V

1. a
2. c
3. c
4. d
5. a
6. b
7. b
8. b
9. c
10. b

CHAPTER VI

1. b
2. c
3. c
4. c
5. a
6. d
7. a
8. c
9. d
10. c

CHAPTER VII

1. b
2. b
3. c
4. b
5. b

CHAPTER VIII

1. b
2. c
3. c
4. a
5. c
6. c
7. d
8. d

CHAPTER IX

1. c
2. d
3. b
4. b
5. b
6. a
7. a
8. c

Answers to Exercises

CHAPTER X

1. a
2. d
3. b
4. c
5. d
6. d
7. c
8. b
9. a
10. c

CHAPTER XI

1. a
2. b
3. c
4. b
5. d
6. b
7. a
8. b
9. c
10. a

CHAPTER XII

1. c
2. b
3. c
4. b
5. a
6. d
7. c
8. a
9. b
10. a

Join Our Community

Medical Creations is an educational company focused on providing study tools for Healthcare students.

We want to be as close as possible to our customers, that's why we are active on all the main Social Media platforms.

You can find us here:

www.facebook.com/medicalcreations

www.instagram.com/medicalcreations

www.twitter.com/medicalcreation (no 's')

www.pinterest.com/medicalcreations

Medical Creations

Kindle MatchBook

Kindle MatchBook is a feature that allows customers who have previously purchased a physical book from Amazon.com to receive the Kindle version for a discounted price or even for free.

You can receive a copy of our Medical Terminology Kindle Edition for FREE.

Just go to the Kindle Version page of the book on Amazon and download the ebook.

Read your ebook on any device (phone, tablet, laptop).

Check Out Our Other Books

Medical Terminology: The Best and Most Effective Way to Memorize, Pronounce and Understand Medical Terms

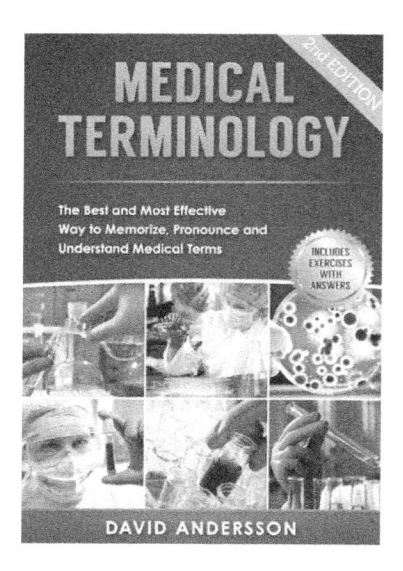

Check Out Our Other Books

EKG/ECG Interpretation: Everything you Need to Know about the 12-Lead ECG/EKG Interpretation and How to Diagnose and Treat Arrhythmias

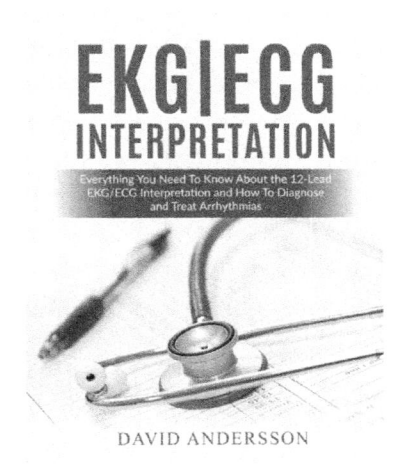

Lab Values: Everything You Need to Know about Laboratory Medicine and its Importance in the Diagnosis of Diseases

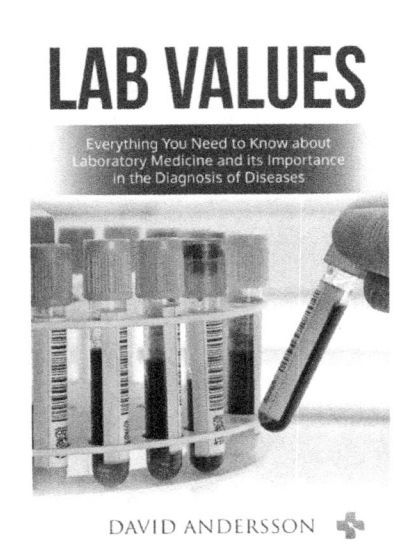

Made in the USA
Middletown, DE
29 August 2020